D0850366

Computational Solution of Nonlinear Operator Equations

Computational Solution of Nonlinear Operator Equations

LOUIS B. RALL

Mathematics Research Center
United States Army
University of Wisconsin

ROBERT E. KRIEGER PUBLISHING COMPANY, INC.

Huntington, New York   1979

Original edition 1969
Reprint 1979 with corrections

Printed and Published by
Robert E. Krieger Publishing Company, Inc.
645 New York Avenue
Huntington, New York 11743

Copyright © 1969 by
John Wiley & Sons, Inc.
Reprinted by Arrangement

All rights reserved. No reproduction in any form of this book,
in whole or in part (except for brief quotation in critical articles
or reviews), may be made without written authorization from
the publisher.

Library of Congress Cataloging in Publication Data
Rall, Louis B.
    Computational solution of nonlinear operator
equations.

    Reprint of the ed. published by Wiley, New
York.
    Includes bibliographies and index.
    1.   Operator equations--Numerical solutions.
2. Nonlinear operators. 3. Iterative methods
(Mathematics)   I. Title.
[QA329.8.R34   1978]   515'.72   78-2378
ISBN 0-88275-667-2

Printed in the United States of America

PREFACE

The applied mathematician and the numerical analyst of today are faced with the problem of finding solutions, or at least approximate solutions of sufficient accuracy, of what may seem to be a bewildering variety of equations: finite and infinite systems, ordinary and partial differential equations subject to initial or boundary conditions, or their combination, and integral or integrodifferential equations. To complicate the matter further many of these equations are nonlinear. From the standpoint of functional analysis, however, all may be formulated in terms of operators that map some linear space into itself, the solutions being sought as elements or points in the corresponding space. Consequently computational methods that work in this general setting for the solution of equations apply to a large number of problems and lead directly to the development of effective and reliable computer programs to obtain accurate approximate solutions to equations in the original or a related space.

This book is written as a text for an advanced numerical analysis or computer science course to introduce these powerful concepts and techniques at the earliest possible stage. The student is assumed to have had basic courses in numerical analysis, computer programming, and linear algebra (of the computational variety) and an introduction to real and complex analysis. The necessary theory from linear and nonlinear functional analysis is developed and illustrated with examples.

This book is an outgrowth of a series of lectures on Newton's method and its applications that I gave at the Mathematics Research Center, United States Army, University of Wisconsin. The support and encouragement provided by Professor J. Barkley Rosser, Director of the Center, and Professor S. C. Kleene, Acting Director during 1966–1967, are gratefully acknowledged. I also presented some of the topics in this volume at the Georgia Institute of Technology in special courses conducted by Professor

M. Zuhair Nashed and at North American Rockwell, Inc., under the auspices of Dr. Jack Zimmerman. These opportunities to test the ideas in this book are sincerely appreciated.

Above all I wish to acknowledge the invaluable assistance of Professor Ramon E. Moore of the University of Wisconsin, who presented the material in this book in nearly final form in a one-semester advanced numerical analysis course. The suggestions made by Professor Moore and his students improved a number of sections. The defects remaining are my sole responsibility. Professor Moore supplemented the course with a lecture on the Lebesgue integral, which he generously contributed as an appendix.

Among the many who have helped with this work I wish to thank in particular Lydia R. Lohr and P. M. Anselone for scientific collaboration, Julia H. Gray, Allen Reiter, and Donald VanEgeren for programming assistance, and Mrs. Shirley White and my mother, Mrs. Jennie Mae Rall, for their efforts in the preparation of the lecture notes and the final manuscript.

LOUIS B. RALL

Madison, Wisconsin

CONTENTS

Computational Solution of
Nonlinear Operator Equations

1

LINEAR SPACES,
OPERATORS, AND EQUATIONS

In ordinary analysis we work with the real or complex number system. Functional analysis is based on the use of linear spaces, which are generalizations of these number systems. The power of functional analysis to deal with a wide variety of problems stems from the fact that linear spaces may be composed of such interesting mathematical objects as vectors with a finite or infinite number of components or functions that satisfy given conditions. Many important transformations and equations in ordinary algebra and analysis may be formulated in terms of linear operators in such spaces. Linear functional analysis, which is the theory of linear spaces, operators, and equations has been developed into an extensive mathematical discipline, with a multitude of applications [1]. (References and notes cited will be found at the end of each chapter.) An introduction to the basic portions of this theory which are essential to the subsequent consideration of nonlinear operators and equations is given in the following. More detailed treatments of linear functional analysis may be found in standard texts on the subject [2, 3, 4].

1. LINEAR SPACES

In abstract terms, a linear vector space is a set of elements, sometimes referred to as *points* or *vectors*, with two operations, called *addition* and *scalar multiplication*, respectively, which satisfy certain conditions. In particular, addition is required to have the same algebraic properties as the addition of real or complex numbers.

1

As a simple example of a linear space, consider the set of two-component vectors

$$R^2 = \{x: x = (\xi_1, \xi_2)\}, \tag{1.1}$$

where ξ_1 and ξ_2 are real numbers. If addition of $x = (\xi_1, \xi_2)$ and $y = (\eta_1, \eta_2)$ is defined by

$$x + y = (\xi_1 + \xi_2, \eta_1 + \eta_2), \tag{1.2}$$

then it follows immediately that this operation has these properties:

A1. *Closure.* If x and y are elements of the space, then $z = x + y$ is also an element of the space.

A2. Addition is *commutative*,

$$x + y = y + x, \tag{1.3}$$

and *associative*,

$$x + (y + z) = (x + y) + z. \tag{1.4}$$

A3. The space has an element 0 which is an *identity element* for addition; that is,

$$x + 0 = x \tag{1.5}$$

for any element x of the space.

In the case of R^2 it is easy to see that

$$0 = (0, 0). \tag{1.6}$$

The element 0 of a linear space is frequently referred to as the *zero vector* or the *origin* of the space.

A4. Each element x of the space has a *negative* $-x$ such that

$$x + (-x) = 0. \tag{1.7}$$

It follows directly from (1.7) that $-0 = 0$ and that the equation

$$x + z = y + z \tag{1.8}$$

for x has the unique solution $x = y$. For convenience in notation it is usual to define the operation of *subtraction* in a linear space by means of the equation

$$x - y = x + (-y) \tag{1.9}$$

for any elements x, y of the space. In R^2, if $x = (\xi_1, \xi_2)$, then

$$-x = (-\xi_1, -\xi_2). \tag{1.10}$$

In any linear space the operation of addition must satisfy **A1** through **A4**. These properties also hold for the real and complex number systems. In brief and abstract terms, any set in which **A1** through **A4** hold is called a *commutative* (or *Abelian*) *group* with respect to the operation of addition.

With each linear space X there is associated a *scalar field* Λ, which is either the field R of real numbers or the field C of complex numbers. A number belonging to Λ is called a *scalar*, and X is said to be a *real* or *complex* linear space, according as $\Lambda = R$ or $\Lambda = C$.

Scalar multiplication is an operation defined between a scalar and an element of the linear space which is required to satisfy certain conditions in order for it to be a generalization of the operation of multiplication of one real or complex number by another; for example, if we introduce scalar multiplication into R^2 by the definition

$$\lambda x = (\lambda \xi_1, \lambda \xi_2) \tag{1.11}$$

for real λ and $x = (\xi_1, \xi_2)$ in R^2, it is evident that this operation has the following properties:

M1.   *Closure.*   If λ is any scalar and x is any element of the space, then λx is an element of the space.

M2.   *Consistency with addition.*   For any element x of the space

$$\begin{aligned} 1 \cdot x &= x, \\ 0 \cdot x &= 0, \\ (-1)x &= -x. \end{aligned} \tag{1.12}$$

M3.   *Distribution across scalar and vector addition.*   For any scalars λ, μ and elements x, y of the space

$$\begin{aligned} (\lambda + \mu)x &= \lambda x + \mu x, \\ \lambda(x + y) &= \lambda x + \lambda y. \end{aligned} \tag{1.13}$$

In any linear space scalar multiplication is required to satisfy conditions **M1** through **M3**. At this point it is possible to collect the foregoing observations to give a precise definition of the mathematical concept of a linear space.

Definition 1.1.   A set X, together with operations called addition, which satisfies **A1** through **A4**, and scalar multiplication, which satisfies **M1** through **M3**, is called a *linear space* over the scalar field Λ.

An example of a real linear space, namely R^2, has already been given. Actually R and C are also extremely simple linear spaces. Another example of an important real linear space is furnished by the set $C[0, 1]$ of real functions $x = x(s)$ which are continuous on the closed interval $0 \leq s \leq 1$, with addition and scalar multiplication defined in the natural way. Since the sum of two continuous functions is continuous, $z = x + y$ or

$$z(s) = x(s) + y(s), \qquad 0 \le s \le 1, \tag{1.14}$$

is an element of $C[0, 1]$ if $x = x(s)$ and $y = y(s)$ are. Thus addition satisfies **A1**, and since (1.14) expresses a relationship between real numbers for each s in the interval $[0, 1]$, conditions **A2** through **A4** are also obviously satisfied. Similarly, for any real λ,

$$\lambda x = \lambda x(s), \qquad 0 \le s \le 1, \tag{1.15}$$

will be an element of $C[0, 1]$ for any $x = x(s)$ belonging to it, and it is easy to see that conditions **M2** through **M3** are also satisfied for scalar multiplication defined in this way.

From a conceptual standpoint the vectors $x = (\xi_1, \xi_2)$ which form the space R^2 and the functions $x = x(s)$ belonging to $C[0, 1]$ are not as different as they may seem at first. The vector may be regarded as being a table of values for a function $x = x(i)$ which is defined only on the set of *coordinates* $i = 1, 2$, with

$$x(i) = \xi_i, \qquad i = 1, 2, \tag{1.16}$$

being the *components* of the vector x. On the other hand, the function $x = x(s)$ could be regarded as being a vector for which each number s in the interval $0 \le s \le 1$ is a coordinate, the real number $x(s)$ being the component of x at the coordinate s. Thus $x = x(s)$ could be considered to be a vector with infinitely many components.

EXERCISES 1

1. For each of the following sets, frame definitions of addition and scalar multiplication for which they will form linear spaces over the given scalar field.
   (a) The set R^n of vectors $x = (\xi_1, \xi_2, \ldots, \xi_n)$, where n is a fixed positive integer and ξ_i is real, $i = 1, 2, \ldots, n$. $\Lambda = R$.
   (b) The set C^n of vectors $z = (\zeta_1, \zeta_2, \ldots, \zeta_n)$ with complex components $\zeta_i, i = 1, 2, \ldots n$. $\Lambda = C$.
   (c) The set $A(D)$ of complex functions $f = f(\zeta)$ which are analytic in the interior of the unit disk $|\zeta| \le 1$. $\Lambda = C$.
   (d) The set HF of infinite real sequences $x = (\xi_1, \xi_2, \ldots)$ having the property that only a finite number of their components are nonzero. $\Lambda = R$.
   (e) The set Π of all polynomials with complex coefficients. $\Lambda = C$.
2. Give a geometric interpretation of the space R^2 and geometric constructions for the operations of addition and scalar multiplication.
3. Show that any polynomial of degree less than or equal to 2 with real coefficients corresponds to a unique vector in R^3, where

$$R^3 = \{x : x = (\xi_1, \xi_2, \xi_3)\}$$

and the ξ, $i = 1, 2, 3$ are real numbers. Conversely, show that any vector in R^3 defines a unique polynomial in this class.

2. PRODUCT SPACES

The definitions (1.2) and (1.14) of addition, and (1.11) and (1.15) of scalar multiplication used in the previous section are special examples of a general technique which can be used to form linear spaces. The definition (1.14), in particular, related the addition of vectors to the addition of real numbers. Since the addition of real numbers satisfies the conditions **A1** through **A4**, these properties are transferred to the vector addition operation by its definition.

Similarly, definition (1.15) related the scalar multiplication of functions to the multiplication of real numbers, and thus obtained immediate satisfaction of conditions **M1** through **M3**.

More generally, suppose that X_1 and X_2 are linear spaces over the same scalar field Λ. Their (*cartesian*) *product* $X_1 \times X_2$ is the set of ordered pairs

$$X_1 \times X_2 = \{(x_1, x_2) : x_1 \in X_1, x_2 \in X_2\}. \tag{2.1}$$

(The notation $x_i \in X_i$ means that x_i is an element of the space X_i, $i = 1, 2$.) Denoting $X_1 \times X_2$ by X, the set X is composed of vectors $x = (x_1, x_2)$, the first components of which are elements of X_1 and the second components are drawn from X_2.

The *natural* definitions of addition and scalar multiplication for X are

$$x + y = (x_1 + y_1, x_2 + y_2), \tag{2.2}$$
$$\lambda x = (\lambda x_1, \lambda x_2), \tag{2.3}$$

respectively, for $x = (x_1, x_2)$, $y = (y_1, y_2)$ in X and $\lambda \in \Lambda$.

Since X_1 and X_2 are linear spaces over Λ, the properties **A1** through **A4** of addition are transferred to addition in X by (2.2), and the properties **M1** through **M3** of scalar multiplication are transferred from X_1 and X_2 to X by (2.3). This proves the following theorem.

Theorem 2.1. The cartesian product of two linear spaces over the same scalar field Λ is a linear space over Λ for the natural definitions of addition and scalar multiplication.

Since the spaces X_1 and X_2 do not need to be related beyond having the same scalar field, a number of mathematical problems can be formulated as a single equation in a product space; for example, consider the ordinary differential equation

$$\frac{dy}{dt} = f(t, y), \qquad 0 \le t \le T, \tag{2.4}$$

for some given $T > 0$, together with the initial condition

$$y(0) = y_0. \tag{2.5}$$

If $f(t, y)$ is real and continuous, it is natural to seek a solution $y = y(t)$ of (2.4) in the space $C'[0, T]$ of all real functions which have continuous first derivatives for $0 \le t \le T$. (The notation $C^{(n)}[a, b]$ for the space of real functions $y = y(t)$ with continuous nth derivatives on the interval $a \le t \le b$ will be used throughout, $n = 0, 1, 2, \ldots$, with $C^{(0)}[a, b] = C[a, b]$.)

For any $y = y(t)$ in $C'[0, T]$ the notation $\delta_0 y$ will be defined by the equation

$$\delta_0 y = y(0), \tag{2.6}$$

which can be interpreted in the following way: An *operator*, denoted by δ_0, is applied to the function $y(t)$ in the space to obtain its value at $t = 0$ as a result of the operation. Using this notation, (2.4) and (2.5) may be replaced by the single equation

$$\left(\frac{dy}{dt} - f(t, y), \delta_0 y - y_0 \right) = (0, 0) \tag{2.7}$$

in $C[0, T] \times R$.

Operators such as δ_0, which map from a linear space into its scalar field, are called *functionals*. It was the study of such operators that gave functional analysis its name [5].

The natural definitions of addition and scalar multiplication, and the conclusions of Theorem 2.1 may be extended to spaces which are the cartesian products of more than two linear spaces over the same scalar field.

Given a finite number of spaces $X_1, X_2, \ldots, X_n$, the elements x of the product space

$$X = \prod_{i=1}^{n} X_i \tag{2.8}$$

would usually be written in the vector notation

$$x = (x_1, x_2, \ldots, x_n), \qquad x_i \in X_i, \qquad i = 1, 2, \ldots, n. \tag{2.9}$$

Thus the vector x would have components that are themselves elements of linear spaces.

It is also customary to use vector notation in the case of a countably infinite set of linear spaces $X_1, X_2, \ldots$. Here

$$x = (x_1, x_2, \ldots), \qquad x_i \in X_i, \qquad i = 1, 2, \ldots, \tag{2.10}$$

would denote an element of the product space

$$X = \prod_{i=1}^{\infty} X_i. \tag{2.11}$$

To express an element of the product of an uncountable infinity of linear spaces, it is convenient to use functional notation; for example, if a linear space X_s is given for each value of s in the interval $0 \le s \le 1$, then we could define an element $x = x(s)$ of the product space

$$X = \prod_{0 \le s \le 1} X_s \tag{2.12}$$

to be a function such that $x(s) \in X_s$ for each s, $0 \le s \le 1$. In general, for any set S of elements s, to each of which there corresponds a linear space X_s, an element $x = x(s)$ of the product

$$X = \prod_{s \in S} X_s \tag{2.13}$$

would be a function on S which takes on the value $x(s) \in X_s$ for each $s \in S$.

As an example of a space of the type (2.12), the space $F[0, 1]$ of *all* real functions $x(s)$ on the interval $0 \le s \le 1$ is the product space

$$F[0, 1] = \prod_{0 \le s \le 1} R. \tag{2.14}$$

This, of course, is a special case of (2.12) for which $X_s = R$.

The natural definitions of addition and scalar multiplication in the space (2.13) would be of $z = x + y$ by

$$z(s) = x(s) + y(s), \qquad s \in S, \tag{2.15}$$

and λx by

$$\lambda x = \lambda x(s), \qquad s \in S, \tag{2.16}$$

respectively. In dealing with products of spaces, it is always assumed that the spaces have a common scalar field.

EXERCISES 2

1. Define the product set $S \times T$, where S is a subset of a space X and T is a subset of a space Y. Under what conditions will $S \times T$ be a linear subspace of $X \times Y$?
2. Formulate the differential equation

$$\frac{d^2 y}{dt^2} = f\left(t, y, \frac{dy}{dt^2}\right), \qquad 0 \le t \le T,$$

together with the initial conditions

$$y(0) = y_0, \qquad \dot{y}(0) = \dot{y}_0,$$

($\dot{y} = dy/dt$) as a single equation in the space

$$X = C''[0, T] \times C'[0, T] \times R \times R.$$

Hint. Set $z(t) = dy/dt$, and use the functional δ_0 defined by (2.6).

3. State and prove a generalization of Theorem 2.1 for the space X defined by (2.13), together with the definitions (2.15) and (2.16). *Hint.* Show that the violation of any of the conditions **A1** through **A4** or **M1** through **M3** contradicts the assumption that X_s is a linear space for at least one $s \in S$.

3. LINEAR DEPENDENCE AND DIMENSIONALITY

If $\{x_1, x_2, \ldots, x_n\}$ is a finite set of vectors belonging to some linear space X, then any vector x of the form

$$x = \sum_{i=1}^{n} \alpha_i x_i, \tag{3.1}$$

where $\alpha_1, \alpha_2, \ldots, \alpha_n$ are scalars, is called a *linear combination* of $x_1, x_2, \ldots,$ x_n. The totality of vectors of the form (3.1) is called the *span* of $x_1, x_2, \ldots, x_n$. It is readily verified that the span of a finite set of vectors is a linear space which is a *subspace* of the original space X, that is, it is a linear space contained in X.

If the vectors $x_1, x_2, \ldots, x_n$ are such that

$$\sum_{i=1}^{n} \alpha_i x_i = 0 \tag{3.2}$$

if and only if $\alpha_1 = \alpha_2 = \cdots = \alpha_n = 0$, then $x_1, x_2, \ldots, x_n$ are said to be *linearly independent*. If (3.2) is satisfied with at least one α_i nonzero, then the vectors $x_1, x_2, \ldots, x_n$ are said to be *linearly dependent*, and it follows that at least one of them is a linear combination of the others. A finite set of linearly independent vectors is called a *basis* for its span. A linear vector space is said to be *n-dimensional* if it has a basis consisting of n nonzero vectors, n being a positive integer. Otherwise it is said to be *infinite-dimensional*, unless it is the space consisting only of the zero vector 0, which will be defined to be *zero-dimensional*.

If X is n-dimensional, it can be regarded as being equivalent to R^n or C^n, once a basis $x_1, x_2, \ldots, x_n$ for X has been chosen [2]. Here for

$$x = \alpha_1 x_1 + \alpha_2 x_2 + \cdots + \alpha_n x_n, \tag{3.3}$$

we would establish the correspondence

$$x \longleftrightarrow (\alpha_1, \alpha_2, \ldots, \alpha_n), \tag{3.4}$$

with $(\alpha_1, \alpha_2, \ldots, \alpha_n)$ in R^n or C^n according as to whether X is a real or complex linear space.

Given the basis $x_1, x_2, \ldots, x_n$ for the space X, the process of forming the vector $(\alpha_1, \alpha_2, \ldots, \alpha_n)$ is called *analysis* of x in terms of the basis, and the formation of x by (3.3) from the *coefficient vector* $(\alpha_1, \alpha_2, \ldots, \alpha_n)$ is called *synthesis* of x.

The representation in (3.3) is unique, for suppose also that

$$x = \beta_1 x_1 + \beta_2 x_2 + \cdots + \beta_n x_n. \tag{3.5}$$

Subtracting (3.5) from (3.3) gives

$$(\alpha_1 - \beta_1)x_1 + (\alpha_2 - \beta_2)x_2 + \cdots + (\alpha_n - \beta_n)x_n = 0, \tag{3.6}$$

from which

$$\alpha_i = \beta_i, \qquad i = 1, 2, \ldots, n, \tag{3.7}$$

since $x_1, x_2, \ldots, x_n$ are assumed to be linearly independent.

The concepts of a basis as well as analysis and synthesis may be extended to certain infinite-dimensional spaces of practical importance [6]. In this case a basis is a set of vectors such that any element of the space can be represented uniquely as a linear combination, possibly infinite, of vectors belonging to the set; for example, the set of complex functions

$$f_n(\zeta) = \zeta^n, \qquad n = 0, 1, 2, \ldots, \tag{3.8}$$

is a basis for the space $A(D)$ of complex functions $f = f(\zeta)$ which are analytic in the unit disk $|\zeta| < 1$. Analysis and synthesis in this space are based on the power series expansion

$$f(\zeta) = \alpha_0 + \alpha_1 \zeta + \alpha_2 \zeta^2 + \cdots. \tag{3.9}$$

By (3.9), $A(D)$ generates a space of infinite complex vectors

$$a = (\alpha_0, \alpha_1, \alpha_2, \ldots)$$

which is a subspace of C^∞, the space of all infinite complex vectors. Since there are elements of C^∞ for which the expansion (3.9) does not yield an element of $A(D)$, the subspace of C^∞ generated by $A(D)$ is what is called a *proper* subspace of C^∞, that is, it is not all of C^∞.

In principle, analysis and synthesis of elements of a linear space with respect to a basis for the space are exact operations. Actual computation, however, is limited to the use of finite sets of numbers, which makes it impossible, in general, to give exact representations of elements of infinite-dimensional spaces such as $A(D)$. A central problem area of computational mathematics is concerned with finding suitable finite representations of elements of infinite-dimensional spaces and the interpretation of the results of finite computations in such spaces.

One method of obtaining an element of a finite-dimensional space which corresponds to a given element of an infinite-dimensional space is called *projection*. A precise definition of this operation will be given later; it may be illustrated by the use of the complex polynomial

$$p_n(\zeta) = \alpha_0 + \alpha_1 \zeta + \cdots + \alpha_n \zeta^n \tag{3.10}$$

to correspond to the element $f(\zeta)$ of $A(D)$ given by (3.9). It is evident that many distinct elements of $A(D)$ will give rise to a single polynomial of the form (3.10). The linear space of all polynomials (3.10) for fixed n is an $(n + 1)$-dimensional subspace of $A(D)$ which is equivalent to the space C^{n+1} of complex vectors $a = (\alpha_0, \alpha_1, \ldots, \alpha_n)$. Conversely, given a vector $a \in C^{n+1}$, we may construct the polynomial $p_n(\zeta)$, which will be an *approximation* to the element $f(\zeta)$ of $A(D)$, in that it is a partial sum of the power series (3.9). This will, of course, be an exact representation if $f(\zeta)$ is actually a polynomial of degree n or less. This process of constructing elements of $A(D)$ from vectors in C^{n+1} is sometimes called *injection* of the space C^{n+1} into the space $A(D)$.

Another process frequently used in numerical analysis is commonly known as *discretization*. To illustrate this, suppose that the infinite-dimensional space under consideration is the space $C[0, 1]$ of continuous real functions $x = x(s)$ on $0 \leq s \leq 1$. From the interval $[0, 1]$ we may choose n points,

$$0 \leq s_1 < s_2 < \cdots < s_n \leq 1, \tag{3.11}$$

which form a set called a *partition* of the interval. For a given function $x(s)$, the values

$$x(s_i) = \xi_i, \qquad i = 1, 2, \ldots, n, \tag{3.12}$$

form a vector $(\xi_1, \xi_2, \ldots, \xi_n)$ which is an element of R^n. The inverse process of constructing a continuous function $x(s)$ for which (3.12) holds from a given vector $(\xi_1, \xi_2, \ldots, \xi_n)$ is usually called *interpolation*. There are many methods of interpolation, most of which are based on the use (explicitly or implicitly) of a set of continuous *interpolating functions* $\phi_1(s), \phi_2(s), \ldots, \phi_n(s)$ such that

$$\phi_i(s_j) = \delta_{ij} = \begin{cases} 0, & i \neq j, \\ 1, & i = j, \end{cases} \tag{3.13}$$

$$i, j = 1, 2, \ldots, n.$$

(The symbol δ_{ij} is called the *Kronecker delta* and is used frequently.)

It is easy to show on the basis of (3.13) that the functions $\phi_i(s)$, $i = 1, 2, \ldots, n$, are linearly independent and, hence, form a basis for an n-dimensional subspace $C_n[0, 1]$ of $C[0, 1]$. The function

$$\phi(s) = \sum_{i=1}^{n} \xi_i \, \phi_i(s) \tag{3.14}$$

is an approximation to $x(s)$ in the sense that $\phi(s_i) = x(s_i) = \xi_i, i = 1, 2, \ldots, n$.

The term *approximation* has been used in a general way. To frame a precise definition we must be able to specify a measure of accuracy (or error). The necessary concepts will be introduced in the next section.

EXERCISES 3

1. Show that every set of vectors that contains the zero vector is linearly dependent. In an n-dimensional space show that every set of vectors containing $n + 1$ or more vectors is linearly dependent.

2. Show that

$$x_1 = (4, -1, 1),$$
$$x_2 = (1, -2, 3),$$
$$x_3 = (-4, -1, 2)$$

form a basis for R^3. Develop a formula for the analysis of an arbitrary vector $x = (\xi_1, \xi_2, \xi_3)$ in terms of this basis.

3. Show that if

$$\sum_{n=0}^{\infty} |\alpha_n| < \infty \tag{I}$$

the function $f(\zeta)$ obtained from the vector $a = (\alpha_0, \alpha_1, \alpha_2, \ldots)$ by (3.9) is an element $A(D)$. Find an element of $A(D)$ for which (I) is not satisfied. Hence (I) is only a sufficient condition for a vector a to represent an element $f(\zeta)$ of $A(D)$.

4. In an infinite-dimensional space X a set of vectors is called a *Hamel basis* for X if any vector $x \in X$ can be represented as a finite linear combination of vectors in the set [2]. Exhibit a Hamel basis for the space HF defined in Problem 1d, Exercises 1.

5. For the partition (3.11) of $[0, 1]$ show that the *Lagrange interpolating polynomials*

$$\phi_i(s) = \prod_{\substack{j=1 \\ j \neq i}}^{n} \frac{s - s_j}{s_i - s_j}, \qquad i = 1, 2, \ldots, n,$$

satisfy conditions (3.13) and that for these functions (3.14) defines the unique polynomial of degree $n - 1$ (or lower) such that

$$\phi(s_i) = \xi_i, \qquad i = 1, 2, \ldots, n.$$

Much of classical numerical analysis is based on the approximation of functions by polynomials of this type.

6. In the partition (3.11) of $[0, 1]$ assume for simplicity that $s_1 = 0$, $s_n = 1$. Consider the functions

$$\phi_1(s) = \begin{cases} \dfrac{s_2 - s}{s_2}, & 0 \leq s \leq s_2, \\ 0, & s_2 \leq s \leq 1, \end{cases}$$

$$\begin{aligned} \phi_i(s) &= 0, & 0 \leq s \leq s_{i-1}, \\ &= \frac{s - s_{i-1}}{s_i - s_{i-1}}, & s_{i-1} \leq s \leq s_i, \\ &= \frac{s_{i+1} - s}{s_{i+1} - s_i}, & s_i \leq s \leq s_{i+1}, \\ &= 0, & s_{i+1} \leq s \leq 1, \end{aligned}$$

for $i = 2, 3, \ldots, n - 1$.

$$\phi_n(s) = \begin{cases} 0, & 0 \leq s \leq s_{n-1}, \\ \dfrac{s - s_{n-1}}{1 - s_{n-1}}, & s_{n-1} \leq s \leq 1. \end{cases}$$

Show that $\phi_1(s)$, $\phi_2(s)$, $\ldots$, $\phi_n(s)$ are continuous functions on $[0, 1]$ which satisfy (3.13). What is the form of the function $\phi(s)$ given by (3.14) in this case? For $n = 4$, choose a partition $0 = s_1 < s_2 < s_3 < s_4 = 1$ of $[0, 1]$ and draw graphs of the functions $\phi_i(s)$, $i = 1, 2, 3, 4$. Choose some values for the coefficients ξ_i, $i = 1, 2, 3, 4$, and draw a graph of the resulting function $\phi(s)$.

The interpolating functions introduced in this problem, and the interpolating polynomial discussed in Problem 5 are both special cases of what are called *spline functions* [7], which have many interesting properties and important applications in numerical analysis.

7. Show that the solutions of the ordinary differential equation

$$\frac{d^n y}{dx^n} + c_1 \frac{d^{n-1} y}{dx^{n-1}} + \cdots + c_{n-1} \frac{dy}{dx} + c_n y = 0$$

form an n-dimensional subspace of $C^\infty(R)$, the space of real functions that have continuous derivatives of all orders.

4. NORMED LINEAR SPACES

The algebraic structure of a linear space is similar to that of the real or complex number system, which permits the extension of many algebraic techniques to problems in a more general setting. However, to deal with other concepts of theoretical and computational importance, such as accuracy of approximation, convergence of sequences or series, and so on, it is necessary to introduce additional structure into such spaces. The *metric* (or *topological*) structure considered here is based on a simple generalization of the idea of the absolute value of a real number or modulus of a complex number.

Suppose that for each element x of a linear space X, a nonnegative real number $\|x\|$, called the *norm* of x, is defined and the following conditions are satisfied:

N1 $\|x\| > 0$ if $x \neq 0$, $\|0\| = 0$,

N2 $\|\lambda x\| = |\lambda| \cdot \|x\|$ for all $\lambda \in \Lambda$,

N3 $\|x + y\| \leq \|x\| + \|y\|$.

Then X is called a *normed linear space*.

It is easy to see that R (or C) is a normed linear space for $\|x\| = |x|$. From a geometrical standpoint $\|x\|$ may be interpreted as being the distance from the origin 0 of the space to the point x, or as the *length* of the vector x. Condition N2 thus states that scalar multiplication changes the length of a vector by a factor equal to the *magnitude* (absolute value or modulus) of the multiplier, and **N3** is the *triangle inequality*, which asserts that the length of one side of a triangle does not excede the sum of the lengths of the other two sides.

In a normed linear space X the *distance* $d(x, y)$ from a point x to a point y is defined to be

$$d(x, y) = \|x - y\|. \tag{4.1}$$

Consequently, if y is regarded as being an approximation to x, the *error* of the approximation is simply $\|x - y\|$. If $x \neq 0$, the *relative error* is $\|x - y\| / \|x\|$ and the *percentage error* is 100 times the relative error.

In general it is possible to introduce the definition of a norm into a linear space in many different ways; for example, in R^2 we could define

$$\|x\|_1 = |\xi_1| + |\xi_2|, \tag{4.2}$$

$$\|x\|_2 = \sqrt{\xi_1{}^2 + \xi_2{}^2}, \tag{4.3}$$

$$\|x\|_\infty = \max \{|\xi_1|, |\xi_2|\}, \tag{4.4}$$

or, more generally, for $p \geq 1$, by

$$\|x\|_p = (|\xi_1|^p + |\xi_2|^p)^{1/p}. \tag{4.5}$$

For these definitions the satisfaction of conditions **N1** and **N2** is obvious. Condition **N3** is also easily verified in the case of (4.2) and (4.4). For (4.3) we must prove that

$$\sqrt{(\xi_1 + \eta_1)^2 + (\xi_2 + \eta_2)^2} \leq \sqrt{\xi_1{}^2 + \xi_2{}^2} + \sqrt{\eta_1{}^2 + \eta_2{}^2}. \tag{4.6}$$

To do this observe for real ξ, η that

$$0 \leq (|\xi| - |\eta|)^2, \tag{4.7}$$

or

$$2 |\xi\eta| \leq \xi^2 + \eta^2. \tag{4.8}$$

Setting $\xi = \xi_1\eta_2$, $\eta = \xi_2\eta_1$, we have

$$|\xi_1\eta_1|^2 + 2 |\xi_1\eta_2\xi_2\eta_1| + |\xi_2\eta_2|^2$$
$$\leq |\xi_1\eta_1|^2 + |\xi_1\eta_2|^2 + |\xi_2\eta_1|^2 + |\xi_2\eta_2|^2, \tag{4.9}$$

or

$$|\xi_1\eta_1| + |\xi_2\eta_2| \leq \sqrt{|\xi_1\eta_1|^2 + |\xi_1\eta_2|^2 + |\xi_2\eta_1|^2 + |\xi_2\eta_2|^2}. \tag{4.10}$$

Consequently

$$|\xi_1 + \eta_1|^2 + |\xi_2 + \eta_2|^2 \leq \xi_1{}^2 + 2 |\xi_1\eta_1| + \eta_1{}^2 + \xi_2{}^2 + 2 |\xi_2\eta_2| + \eta_2{}^2,$$
$$\leq (\sqrt{\xi_1{}^2 + \xi_2{}^2} + \sqrt{\eta_1{}^2 + \eta_2{}^2})^2, \tag{4.11}$$

which establishes (4.6). The norm $\|x\|_2$ defined by (4.3) is called the *Euclidean norm* for R^2.

For the more general case (4.5) we verify **N3** by proving that

$$\sqrt[p]{|\xi_1 + \eta_1|^p + |\xi_2 + \eta_2|^p} \leq \sqrt[p]{|\xi_1|^p + |\xi_2|^p} + \sqrt[p]{|\eta_1|^p + |\eta_2|^p}, \quad (4.12)$$

which is called *Minkowski's inequality* [4,8]. The norms defined by (4.5) are called *Minkowski norms* [2]. This family of norms actually includes (4.4) as the limiting case as $p \to \infty$. To see this suppose that $\|x\|_\infty = |\xi_1| \neq 0$ and note that

$$\|x\|_p = |\xi_1| \left(1 + \left|\frac{\xi_2}{\xi_1}\right|^p\right)^{1/p}. \quad (4.13)$$

Since $|\xi_2/\xi_1| \leq 1$,

$$\lim_{p \to \infty} \|x\|_p = |\xi_1| = \|x\|_\infty. \quad (4.14)$$

The Minkowski norms for R^2 may be extended to R^n (or C^n) very simply; for example, for $x \in R^n$, we could define

$$\|x\|_p = \left(\sum_{i=1}^n |\xi_i|^p\right)^{1/p}, \quad 1 \leq p \leq \infty, \quad (4.15)$$

with

$$\|x\|_\infty = \max_{(i)} |\xi_i|, \quad i = 1, 2, \ldots, n. \quad (4.16)$$

The space R^n or C^n with the norm $\|x\|_p$ is denoted by $R_p{}^n$ or $C_p{}^n$ respectively.

In infinite-dimensional spaces the introduction of a norm is subject to somewhat more restrictive conditions; for example, the infinite series corresponding to $p = 1$,

$$\|x\|_1 = \sum_{i=1}^\infty |\xi_i| \quad (4.17)$$

will converge only for vectors x belonging to a subset of R^∞. It turns out that this subset, which is denoted by R_1^∞, is a normed linear space with elements belonging to a subspace of R^∞. To show this we need to verify properties **A1** through **A4** and **M1** through **M3** for the set of vectors $x = (\xi_1, \xi_2, \ldots)$ such that

$$\sum_{i=1}^\infty |\xi_i| < \infty, \quad (4.18)$$

with addition and scalar multiplication defined as in R^∞. In the same way, for $1 \leq p < \infty$, we may define R_p^∞ (or C_p^∞) to be the subset of R^∞ (or C^∞) consisting of vectors x such that

$$\|x_p\| = \left(\sum_{i=1}^\infty |\xi_i|^p\right)^{1/p} < \infty. \quad (4.19)$$

To show that these subsets are normed linear spaces for the original definitions of addition and scalar multiplication we must establish Minkowski's inequality for infinite series [8],

$$\left(\sum_{i=1}^{\infty} |\xi_i + \eta_i|^p\right)^{1/p} \le \left(\sum_{i=1}^{\infty} |\xi_i|^p\right)^{1/p} + \left(\sum_{i=1}^{\infty} |\eta_i|^p\right)^{1/p}, \qquad 1 \le p < \infty, \quad (4.20)$$

and verify **A1** through **A4** and **M1** through **M3**. The space R_∞^∞ (or C_∞^∞) is the space of all bounded infinite real (or complex) sequences, with the norm defined to correspond to (4.16).

It is interesting to note that although $R_p{}^n$ and $R_p{}^n$, consist of the same elements for finite n this is not true for R_p^∞ and $R_{p'}^\infty$, if $p' \ne p$; for example, the vector

$$x = \left(1, \frac{1}{2}, \frac{1}{3}, \ldots, \frac{1}{n}, \ldots\right) \qquad (4.21)$$

belongs to R_2^∞, but not to R_1^∞.

For spaces consisting of functions it is natural to replace the operation of summation in the definition of the norm by integration; for example, the subset $F_1[0, 1]$ of the space $F[0, 1]$ of all real functions $x = x(s)$ on $0 \le s \le 1$ for which the Riemann integral

$$\|x\|_1 = \int_0^1 |x(s)| \, ds \qquad (4.22)$$

exists and is finite is a normed linear space, as are the spaces $F_p[0, 1]$ of real functions for which the integrals

$$\|x\|_p = \left(\int_0^1 |x(s)|^p \, ds\right)^{1/p}, \qquad 1 \le p < \infty, \qquad (4.23)$$

exist and are finite. The space $F_\infty[0, 1]$ consists of all bounded real functions with

$$\|x\|_\infty = \sup_{[0,1]} |x(s)|. \qquad (4.24)$$

It is convenient to introduce some geometric terminology to describe some important subsets of normed linear spaces. In a normed linear space X the set

$$U_0 = \{x : \|x\| < 1\} \qquad (4.25)$$

is called the *open unit ball*, the set

$$\bar{U}_0 = \{x : \|x\| \le 1\} \qquad (4.26)$$

is called the *closed unit ball*, and the set

$$S_0 = \{x : \|x\| = 1\} \qquad (4.27)$$

is called the *unit sphere*. Scalar multiplication of the elements of these sets by r and translation by adding x_0 to each of the elements of the resulting sets give:

$$U(x_0, r) = \{x: \|x - x_0\| < r\},$$
$$\bar{U}(x_0, r) = \{x: \|x - x_0\| \le r\}, \qquad (4.28)$$
$$S(x_0, r) = \{x: \|x - x_0\| = r\},$$

which are referred to respectively as the open ball, closed ball, and sphere with *center* x_0 and *radius* r. A subset of a normed linear space X is said to be *bounded* if it is contained in some ball of finite radius.

In a normed linear space it is possible to discuss the analytic notions of convergence and limits of sequence of elements of the space. An element x^* is said to be the *limit* of a sequence $\{x_m\}$ of elements of X if and only if

$$\lim_{m \to \infty} \|x_m - x^*\| = 0. \qquad (4.29)$$

If (4.29) holds, the sequence $\{x_m\}$ is said to be *convergent*, and to converge to x^*. If x^* exists, **N3** may be used to show that it is unique. An infinite series $x_1 + x_2 + \cdots$ of elements of a normed linear space is *convergent* if the sequence $\{x_m\}$ of *partial sums*

$$x_m = \sum_{i=1}^{m} x_i \qquad (4.30)$$

converges. In this case the limit x^* of $\{x_m\}$ is called the *sum* of the infinite series.

If $\|x\|$ and $\|x\|'$ are two different norms for a linear space, they are said to be *topologically equivalent* if real constants a, b exist such that $b > a > 0$ and

$$a \|x\| \le \|x\|' \le b \|x\| \qquad (4.31)$$

for all $x \in X$ [2]. This definition is symmetric in the sense that if (4.31) holds then for $a' = 1/b$, $b' = 1/a$

$$b' > a' > 0 \qquad (4.32)$$

and

$$a' \|x\|' \le \|x\| \le b' \|x\|'; \qquad (4.33)$$

for example, in R^2 if $\|x\| = \|x\|_\infty$ and $\|x\|' = \|x\|_2'$ we have

$$\|x\|_\infty \le \|x\|_2 \le \sqrt{2} \|x\|_\infty. \qquad (4.34)$$

If $\|x\|$ and $\|x\|'$ are topologically equivalent norms for X, then (4.29) holds if and only if

$$\lim_{m \to \infty} \|x_m - x^*\|' = 0. \qquad (4.35)$$

In a finite-dimensional normed linear space it is known that any two norms are topologically equivalent [2]. In actual computation in R^n it is usually most convenient to use $\|x\|_\infty$ (sometimes called the *max norm*), since it is available with the least amount of calculation. If the constants a and b in (4.31) are known for some other norm $\|x\|'$ of interest, the value found for $\|x\| = \|x\|_\infty$ may provide satisfactory estimates for $\|x\|'$ without further calculation. Of the remaining Minkowski norms $\|x\|_1$ requires the least computational effort, followed by $\|x\|_2$.

If we are given a finite number of normed linear spaces, a norm may be introduced into the product space in much the same way as for a finite dimensional linear space; for example, if $X = X_1 \times X_2$, where $\|x_i\|_{(i)}$ denotes the norm of an element $x_i \in X_i$, $i = 1, 2$, we could define

$$\|x\|_\infty = \max\{\|x_1\|_{(1)}, \|x_2\|_{(2)}\} \tag{4.36}$$

or, for $1 \leq p < \infty$,

$$\|x\|_p = (\|x_1\|_{(1)}^p + \|x_2\|_{(2)}^p)^{1/p}. \tag{4.37}$$

For infinite products of normed linear spaces questions of convergence or integrability arise in connection with the introduction of a norm into the product space. These questions and the method for their resolution are much the same as indicated above for infinite-dimensional vector and function spaces.

EXERCISES 4

1. Graph the unit sphere (circle) in R_1^2, R_2^2, R_3^2, R_∞^2. Calculate the constants a, b in (4.31) as $\|x\|$ and $\|x\|'$ are selected from $\|x\|_1$, $\|x\|_2$, $\|x\|_3$, $\|x\|_\infty$ for R^2. Give a geometric interpretation of inequality (4.31).

2. Prove the Minkowski inequality (4.20) on the basis of the Hölder inequality [4, 8]

$$\sum_{i=1}^n |a_i b_i| \leq \left(\sum_{i=1}^n |a_i|^p\right)^{1/p} \left(\sum_{i=1}^n |b_i|^q\right)^{1/q},$$

where $p > 1$, $1/p + 1/q = 1$. Indicate an extension of Hölder's inequality to infinite series and integrals.

3. Describe the closed unit balls in R_1^3, R_2^3, R_∞^3 in ordinary geometric terminology.

4. For $x_m = (\xi_{m1}, \xi_{m2}, \ldots, \xi_{mn})$, $x^* = (\xi_1^*, \xi_2^*, \ldots, \xi_n^*)$ in R^n, prove that (4.29) holds if and only if

$$\lim_{m \to \infty} |\xi_{mi} - \xi_i^*| = 0, \qquad i = 1, 2, \ldots, n.$$

5. Show that the space $C[0, 1]$ is a normed linear space for

$$\|x\|_\infty = \max_{[0,1]} |x(s)|$$

or for

$$\|x\|_2 = \left(\int_0^1 |x(s)|^2 \, ds\right)^{1/2}.$$

6. Show that the space *HF* of Problem 1*d*, Exercises 1, is a normed linear space for

$$\|x\|_\infty = \max_{(i)}|\xi_i|,$$

$$\|x\|_1 = \sum_{i=1}^{\infty}|\xi_i|,$$

or, if $1 < p < \infty$, for

$$\|x\|_p = \left(\sum_{i=1}^{\infty}|\xi_i|^p\right)^{1/p}.$$

7. Show that all norms of the form (4.37) for $X = X_1 \times X_2$ are topologically equivalent. *Hint.* Interpret $(\|x\|_{(1)}, \|x_2\|_{(2)})$ as a vector in R^2.

5. BANACH SPACES

This important class of normed linear spaces is named after Stefan Banach (1892–1945), a Polish mathematician who made fundamental contributions to linear functional analysis [9]. In the solution of many problems the basic issue at stake is the *existence* of a limit x^* of an infinite sequence $\{x_m\}$ of elements of a normed linear space X. This situation is a familiar one in classical analysis; for example, consider the sequence $\{x_m\}$ of rational numbers defined by

$$x_0 = 1, \qquad x_m = \frac{1}{2}\left(x_{m-1} + \frac{3}{x_{m-1}}\right), \qquad m = 1, 2, \ldots. \tag{5.1}$$

There is no rational number x^* which can be interpreted as the limit of this sequence in the sense of (4.29) (with $\|x\| = |x|$). However, if (5.1) is viewed as generating a sequence $\{x_m\}$ in R, it has a limit x^* that is a solution of the nonlinear equation

$$x^2 = 3. \tag{5.2}$$

Consequently the space of real numbers has a property with respect to limits which the set of rational numbers does not. This property is now defined precisely in the more abstract setting of a normed linear space.

Definition 5.1. A sequence $\{x_m\}$ of elements of a normed linear space is called a *Cauchy* (or *fundamental*) *sequence* if

$$\lim_{m \to \infty} \lim_{p \to \infty} \|x_{m+p} - x_m\| = 0. \tag{5.3}$$

The sequence in R defined by (5.1) is a Cauchy sequence. In R_1^∞ the sequence defined by

$$x_m = \left(1, \frac{1}{2}, \frac{1}{3}, \ldots, \frac{1}{m}, 0, 0, \ldots\right), \qquad m = 1, 2, \ldots, \tag{5.4}$$

is not a Cauchy sequence, owing to the divergence of the harmonic series

$$1 + \tfrac{1}{2} + \tfrac{1}{3} + \cdots. \tag{5.5}$$

In R_2^∞, however, (5.4) does define a Cauchy sequence, which has the vector

$$x^* = \left(1, \frac{1}{2}, \frac{1}{3}, \ldots, \frac{1}{n}, \frac{1}{n+1}, \ldots\right) \tag{5.6}$$

as its limit. Thus the choice of the definition of the norm may be of critical importance in dealing with infinite processes in infinite-dimensional spaces.

Definition 5.2. A normed linear space X is said to be *complete* if every Cauchy sequence of elements of X converges to a limit which is an element of X.

Definition 5.3. A *Banach space* is a complete normed linear space.

Banach spaces are the abstractions of the real and complex number systems in which it is possible to pose and solve numerous problems in practical analysis from a unified standpoint. The variety of such spaces is illustrated by the following examples, with additional ones being indicated in the exercises.

Theorem 5.1. The spaces $R_p{}^n$ and $C_p{}^n$, $1 \le p \le \infty$, are Banach spaces for any positive integer n.

PROOF. It is evident that if a normed linear space is complete for a given norm it is complete for any norm that is topologically equivalent to the given one. Consequently it is sufficient to establish the assertion of the theorem for R_∞^n and C_∞^n. If $\{x_m = (\xi_{m1}, \xi_{m2}, \ldots, \xi_{mn})\}$ is a Cauchy sequence in one of these spaces, then $\{\xi_{mi}\}$, $i = 1, 2, \ldots, n$, are Cauchy sequences in R (or C) and numbers ξ_i^*, $i = 1, 2, \ldots, n$, exist such that

$$\lim_{m \to \infty} |\xi_{mi} - \xi_i^*| = 0, \qquad i = 1, 2, \ldots, n. \tag{5.7}$$

Consequently $\{x_m\}$ has the limit

$$x^* = (\xi_1^*, \xi_2^*, \ldots, \xi_n^*), \tag{5.8}$$

and R_∞^n (or C_∞^n) is complete, which concludes the proof.

In much the same way it may be shown that R_p^∞ and C_p^∞, $1 \le p \le \infty$, are Banach spaces.

On the other hand, many infinite-dimensional normed linear spaces are not complete; for example, consider the norm

$$\|x\| = \left(\int_0^1 |x(s)|^2 \, ds\right)^{1/2} \tag{5.9}$$

for the space $C[0, 1]$. The sequence $\{x_m = x_m(s)\}$ of continuous functions defined by

$$x_m(s) = \begin{cases} 2^m s^{m+1}, & 0 \le s \le \tfrac{1}{2}, \\ 1 - 2^m(1 - s)^{m+1}, & \tfrac{1}{2} \le s \le 1, \end{cases} \tag{5.10}$$

is a Cauchy sequence for the norm (5.9). It would be reasonable to regard the function

$$x^*(s) = \begin{cases} 0, & 0 \le s < \tfrac{1}{2}, \\ \tfrac{1}{2}, & s = \tfrac{1}{2}, \\ 1, & \tfrac{1}{2} < s \le 1, \end{cases} \tag{5.11}$$

to be the limit of the Cauchy sequence $\{x_m\}$, but $x^* = x^*(s) \notin C[0, 1]$.

In a case such as this, it is usually possible to construct a Banach space which has the original space, or a space which can be identified with the original space, as a proper subspace. This process of *completion* of the space is done in the following manner. First, we form the set $\tilde{X}$ of all Cauchy sequences $\tilde{x} = \{x_m\}$ of elements of X. $\tilde{X}$ is a linear space for the natural definitions of addition and scalar multiplication, the scalar field being the same as for X. Define the symbol $(\tilde{x})$ by

$$(\tilde{x}) = \lim_{m \to \infty} \|x_m\|, \tag{5.12}$$

which exists and is finite by the definition of a Cauchy sequence and the triangle inequality **N3**. We can see that $\tilde{X}$ is not a normed linear space for (5.12), since we can have $(\tilde{x} - \tilde{y}) = 0$ for distinct sequences $\tilde{x}$ and $\tilde{y}$. However, $\tilde{X}$ can be partitioned into *equivalence classes* by the *equivalence relation*

$$\tilde{x} = \tilde{y} \leftrightarrow (\tilde{x} - \tilde{y}) = 0. \tag{5.13}$$

The space of equivalence classes $\hat{x}$ will be denoted by $\hat{X}$. If $\{x_m\} \in \hat{x}$, $\tilde{x} = \{x_m\}$ is called a *representative* of the equivalence class $\hat{x}$. Each element x of the original space can be identified with the Cauchy sequence $\{x, x, x, \ldots\}$ and the equivalence class which contains this sequence. Addition of equivalence classes can be defined by saying that $\hat{z} = \hat{x} + \hat{y}$ is the equivalence class which contains the sum of any two representative sequences $\tilde{x} \in \hat{x}$, $\tilde{y} \in \hat{y}$. Scalar multiplication is defined in a similar fashion, so that $\hat{X}$ becomes a linear space over the same scalar field as for X.

If $\tilde{x} \in \hat{x}$, we may define

$$\|\hat{x}\| = (\tilde{x}), \tag{5.14}$$

which will be independent of the choice of $\tilde{x}$ by (5.13). If $\{\hat{x}_m\}$ is a Cauchy sequence of elements of $\hat{X}$, consider the sequence of representatives $\tilde{x}_m \in \hat{x}_m$, $m = 1, 2, \ldots$, where

$$\tilde{x}_m = \{x_{m1}, x_{m2}, x_{m3}, \ldots\}. \tag{5.15}$$

The *diagonal sequence*

$$\tilde{x}^* = \{x_{11}, x_{22}, x_{33}, \ldots\} \tag{5.16}$$

is a Cauchy sequence and thus belongs to an equivalence class $\hat{x}^*$ in the space $\hat{X}$. For the norm defined by (5.15)

$$\lim_{m \to \infty} \hat{x}_m = \hat{x}^*, \tag{5.17}$$

so that $\hat{X}$ is complete, hence is a Banach space.

Starting from the space $C[0, 1]$ with the norm (5.9), this process leads to a space called $L_2[0, 1]$ if the integral in (5.9) is understood to be a *Lebesgue integral*. The Lebesgue integral is used extensively in functional analysis because of its simple properties, for example, with respect to the interchange of limiting processes. The appendix gives a brief definition of this integral; more details are given elsewhere [3,4]. For the present purposes it is adequate to recall that if a function is Riemann-integrable on an interval it is Lebesgue-integrable and the two integrals are equal. Since most of the functions arising in everyday computational problems are Riemann-integrable, the integrals used here can usually be interpreted in their more elementary sense.

The spaces $L_p[0, 1]$, $1 \leq p \leq \infty$, of functions $x = x(s)$, $0 \leq s \leq 1$, such that the Lebesgue integral

$$\|x\|_p = \left(\int_0^1 |x(s)|^p \, ds \right)^{1/p}, \tag{5.18}$$

exists and is finite are Banach spaces which are important in many applications of functional analysis [4]. Strictly speaking, these spaces are composed of equivalence classes of functions, but their elements may be interpreted in the ordinary way by regarding two functions to be equal if the integral (5.18) of their difference vanishes.

EXERCISES 5

1. In a linear space, the *line segment* $L(x_0, x_1)$ joining the points x_0 and x_1 is the set

$$L(x_0, x_1) = \{x : x = \theta x_1 + (1 - \theta)x_0, 0 \leq \theta \leq 1\}.$$

A set is *convex* if it contains the line segments joining any two of its points. Prove that the open and closed unit balls in a Banach space (in fact, in any normed linear space) are convex.

2. Show that the space $C[0, 1]$ of continuous real functions $x = x(s)$, $0 \leq s \leq 1$, is a Banach space for the norm

$$\|x\| = \max_{[0,1]} |x(s)|.$$

3. Show that the space HF (Problem 1d, Exercises 1) is not complete for the norm

$$\|x\|_2 = \left(\sum_{i=1}^{\infty} |\xi_i|^2 \right)^{1/2},$$

or

$$\|x\|_{\infty} = \max_{(i)} |\xi_i|.$$

Describe the spaces obtained by the completion of HF for each of the norms.

4. Graph the functions defined by (5.10) for $m = 1, 2, 4$.

5. Prove that the spaces $C^{(n)}[0, 1]$ are Banach spaces for the norms

$$\|x\|_{\infty} = \max \{ \|x\|, \|x'\|, \ldots, \|x^{(n)}\| \}$$

and

$$\|x\|_1 = \|x\| + \|x'\| + \cdots + \|x^{(n)}\|,$$

where $\|\ \ \|$ denotes the norm in $C[0, 1]$ (see [9]).

6. HILBERT SPACES

Some special Banach spaces of particular importance are called *Hilbert spaces*, after the German mathematician David Hilbert (1862–1943). These spaces are a very natural generalization of the classical Euclidean spaces of ordinary geometry and vector analysis. On the basis of this generalization, Hilbert was able to develop a powerful theory of linear integral equations and infinite systems of equations [10,11].

In addition to the notion of the length of a vector, the idea of the *direction* of a vector (or, more precisely, of the *angle* between two vectors) is preserved in a Hilbert space. The manner in which this is done is shown by the following illustration in R_2^2. By the definition of the norm in R_2^2, we have for $x = (\xi_1, \xi_2)$, $y = (\eta_1, \eta_2)$,

$$\begin{aligned}
\|x - y\|^2 &= (\xi_1 - \eta_1)^2 + (\xi_2 - \eta_2)^2 \\
&= \xi_1^2 - 2\xi_1\eta_1 + \eta_1^2 + \xi_2^2 - 2\xi_2\eta_2 + \eta_2^2 \\
&= \xi_1^2 + \xi_2^2 + \eta_1^2 + \eta_2^2 - 2(\xi_1\eta_1 + \xi_2\eta_2) \\
&= \|x\|^2 + \|y\|^2 - 2(\xi_1\eta_1 + \xi_2\eta_2).
\end{aligned} \tag{6.1}$$

Using the ordinary geometric interpretation of elements of R_2^2 as plane vectors, it is seen that x, y and $x - y$ form a triangle with sides of lengths $\|x\|$, $\|y\|$, and $\|x - y\|$ respectively (see Figure 6.1). By the law of cosines from elementary trigonometry,

$$\|x - y\|^2 = \|x\|^2 + \|y\|^2 - 2 \|x\| \cdot \|y\| \cos \theta, \tag{6.2}$$

where θ is the angle between the sides of the triangle represented by the vectors x and y. Thus by comparison of (6.1) with (6.2) we define

$$\cos \theta = \frac{\xi_1 \eta_1 + \xi_2 \eta_2}{\|x\| \cdot \|y\|}, \tag{6.3}$$

for x, $y \neq 0$ to be the *cosine of the angle θ between x and y*. Ordinarily we would restrict θ to the interval $0 \leq \theta \leq \pi$ in order to obtain an unambiguous value from (6.3).

The numerator of the right side of (6.3) is a quantity that appears frequently in calculations with vectors, namely, the sum of the products of corresponding components. It will be called the *inner product* of the vectors x, $y \in R_2^2$ and is denoted by $\langle x, y \rangle$, so that

$$\langle x, y \rangle = \xi_1 \eta_1 + \xi_2 \eta_2. \tag{6.4}$$

This is sometimes called the *scalar product* of x and y [6]; however, this terminology will not be used here because of possible confusion with the result of the operation of scalar multiplication.

From (6.4) it follows at once that

$$\langle x, x \rangle = \xi_1^2 + \xi_2^2 = \|x\|^2. \tag{6.5}$$

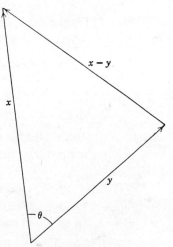

Figure 6.1

To preserve this relationship in the complex space C_2^2, we could define

$$\langle f, z \rangle = \phi_1 \bar{\zeta}_1 + \phi_2 \bar{\zeta}_2, \tag{6.6}$$

for $f = (\phi_1, \phi_2)$, $z = (\zeta_1, \zeta_2)$ in C_2^2, where $\bar{\zeta}_i$ denotes the complex conjugate of ζ_i, $i = 1, 2$. Thus

$$\langle z, z \rangle = \zeta_1 \bar{\zeta}_1 + \zeta_2 \bar{\zeta}_2 = |\zeta_1|^2 + |\zeta_2|^2 = \|z\|^2 \tag{6.7}$$

in C_2^2. It is easy to verify from (6.6) that the inner product has the following properties:

P1 $$\langle f, z \rangle = \overline{\langle z, f \rangle},$$

P2 $$\langle \lambda f + \mu g, z \rangle = \lambda \langle f, z \rangle + \mu \langle g, z \rangle,$$

P3 $$\langle z, z \rangle > 0 \quad \text{if} \quad z \neq 0,$$

for all f, g, z in the space and all scalars λ, μ. Since complex conjugation does not change the value of a real number, the inner product (6.4) of real vectors also satisfies **P1** through **P3**.

Definition 6.1. An *inner product space* is a linear space in which there is defined a scalar-valued inner product $\langle f, z \rangle$ for each f, z in the space, and the inner product satisfies **P1** through **P3**.

Theorem 6.1 (*The Cauchy-Schwarz Inequality*). In an inner product space

$$|\langle f, z \rangle|^2 \leq \langle f, f \rangle \langle z, z \rangle. \tag{6.8}$$

PROOF. For real λ

$$Q(\lambda) = \langle f + \lambda z, f + \lambda z \rangle$$
$$= \lambda^2 \langle z, z \rangle + \lambda(\langle f, z \rangle + \langle z, f \rangle) + \langle f, f \rangle \tag{6.9}$$

is a real quadratic function of the real variable λ of the form

$$Q(\lambda) = a\lambda^2 + b\lambda + c. \tag{6.10}$$

By **P3**, $Q(\lambda) \geq 0$ for all λ, so that the *discriminant* $b^2 - 4ac$ of the quadratic form (6.10) is nonpositive; that is,

$$b^2 - 4ac = (\langle f, z \rangle + \langle z, f \rangle)^2 - 4\langle z, z \rangle \langle f, f \rangle \leq 0. \tag{6.11}$$

(See Figure 6.2, which depicts the case $a \neq 0$.) A scalar multiplier $\phi = e^{i\theta}$ with $|\phi| = 1$ can be found such that

$$\phi\langle f, z \rangle = \langle \phi f, z \rangle = \bar{\phi}\langle z, f \rangle = \langle z, \phi f \rangle \tag{6.12}$$

is real. From (6.11)

$$|\langle \phi f, z \rangle|^2 \leq \langle \phi f, \phi f \rangle \langle z, z \rangle. \tag{6.13}$$

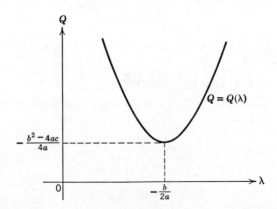

Figure 6.2

By **P1** and **P2**, since $\phi\bar{\phi} = 1$,

$$|\langle f, z \rangle|^2 \leq \langle f, f \rangle \langle z, z \rangle, \tag{6.14}$$

which was to be proved. The definition (6.6) can be extended immediately to $C_2{}^n$ and inequality (6.8) becomes

$$|\langle f, z \rangle| = \left| \sum_{i=1}^{n} \phi_i \bar{\zeta}_i \right| \leq \left(\sum_{i=1}^{n} |\phi_i|^2 \right)^{\frac{1}{2}} \left(\sum_{i=1}^{n} |\zeta_i|^2 \right)^{\frac{1}{2}}. \tag{6.15}$$

Once again multipliers ξ_i can be found with $|\xi_i| = 1$ and such that

$$(\xi_i \phi_i) \bar{\zeta}_i = |\phi_i \zeta_i| \geq 0, \qquad i = 1, 2, \ldots, n. \tag{6.16}$$

As

$$\sum_{i=1}^{n} |\xi_i \phi_i|^2 = \sum_{i=1}^{n} |\phi_i|^2, \tag{6.17}$$

it follows from (6.15) that

$$\sum_{i=1}^{n} |\phi_i \zeta_i| \leq \left(\sum_{i=1}^{n} |\phi_i|^2 \right)^{\frac{1}{2}} \left(\sum_{i=1}^{n} |\zeta_i|^2 \right)^{\frac{1}{2}}, \tag{6.18}$$

which is Holder's inequality with $p = q = 2$ (see Problem 2, Exercises 4).

Corollary 6.1.   An inner product space is a normed linear space for

$$\|z\| = \sqrt{\langle z, z \rangle}. \tag{6.19}$$

PROOF.   Conditions **N1–N2** are obviously satisfied. Condition **N3** follows at once from inequality (6.8).

In a real inner product space the angle θ between the nonzero vectors f and z is defined by

$$\cos \theta = \frac{\langle f, z \rangle}{\sqrt{\langle f, f \rangle \langle z, z \rangle}}, \qquad 0 \leq \theta \leq \pi. \tag{6.20}$$

This definition is always meaningful, in view of inequality (6.8). The Cauchy-Schwarz inequality is thus equivalent to the statement that the magnitude of the cosine of an angle is less than or equal to 1. In spite of its simple nature this inequality is extremely powerful and has been used in the solution of many applied problems. The name of Bunjakovskiĭ is also sometimes associated with this inequality [14]. It should be noted that some authors [6] prefer to define the inner product in a fashion which would give

$$\langle f, z \rangle = \bar{\phi}_1 \zeta_1 + \bar{\phi}_2 \zeta_2 \tag{6.21}$$

in $C_2{}^2$; that is, $\bar{\lambda}$ and $\bar{\mu}$ would appear in **P2** in place of λ and μ respectively. This is most frequent among writers concerned with applications of functional analysis to quantum mechanics. Although this choice does not affect the theory of inner product spaces, it is essential in performing actual computations to be clear as to which definition is being followed.

Definition 6.2.   A *Hilbert space* is an inner product space that is complete for the norm (6.19).

A Hilbert space is thus a special type of Banach space. Although (6.20) does not define an angle in the ordinary sense in a complex Hilbert space, it is still possible to retain the concept of a *right angle*.

Definition 6.3.   Two vectors x, y belonging to a Hilbert space H are said to be *orthogonal* (or *perpendicular*), in symbols, $x \perp y$, if

$$\langle x, y \rangle = 0. \tag{6.22}$$

The zero vector is thus orthogonal to every vector in the space including itself. Two nonzero orthogonal vectors x, y are obviously linearly independent, for if

$$\alpha x + \beta y = 0, \tag{6.23}$$

then

$$\alpha \langle x, x \rangle = \beta \langle y, y \rangle = 0, \tag{6.24}$$

from which it follows that $\alpha = \beta = 0$.

A vector x such that $\|x\| = 1$ is called a *normal* or *unit vector*. A set of vectors $\{x_1, x_2, \ldots\}$ is said to be an *orthonormal* set if

$$\langle x_i, x_j \rangle = \delta_{ij} = \begin{cases} 0, & i \neq j, \\ 1, & i = j, \end{cases} \quad i, j = 1, 2, \ldots. \tag{6.25}$$

The problem of analysis of a vector in terms of an *orthonormal basis* (a basis which is an orthonormal set) has an extremely simple solution. For

$$x = \alpha_1 x_1 + \alpha_2 x_2 + \cdots + \alpha_n x_n + \cdots, \tag{6.26}$$

it follows from (6.25) that

$$\alpha_i = \langle x, x_i \rangle, \quad i = 1, 2, \ldots, \tag{6.27}$$

and

$$\|x\|^2 = \alpha_1{}^2 + \alpha_2{}^2 + \cdots + \alpha_n{}^2 + \cdots. \tag{6.28}$$

An orthonormal basis for an n-dimensional space would, of course, consist of only n vectors; for example,

$$x_1 = (1, 0, 0, \ldots, 0),$$
$$x_2 = (0, 1, 0, \ldots, 0),$$
$$\vdots \tag{6.29}$$
$$x_n = (0, 0, 0, \ldots, 1),$$

is an orthonormal basis for $R_2{}^n$ or $C_2{}^n$. The functions

$$\frac{1}{\sqrt{2\pi}}, \frac{1}{\sqrt{\pi}} \sin s, \frac{1}{\sqrt{\pi}} \cos s, \frac{1}{\sqrt{\pi}} \sin 2s, \frac{1}{\sqrt{\pi}} \cos 2s, \ldots \qquad (6.30)$$

form an orthonormal basis for $L_2[0, 2\pi]$, the space of all real functions which are square-integrable on the interval $0 \leq s \leq 2\pi$ [12], with the inner product defined by

$$\langle x, y \rangle = \int_0^{2\pi} x(s)\, y(s)\, ds. \qquad (6.31)$$

Strictly speaking, the integral in (6.31) is a Lebesgue integral. The analysis of an element $x \in L_2[0, 2\pi]$ in terms of this orthonormal basis yields the familiar *Fourier series*

$$x(s) = \tfrac{1}{2}a_0 + \sum_{n=1}^{\infty} (a_n \cos ns + b_n \sin ns), \qquad (6.32)$$

with

$$a_0 = \frac{1}{\pi} \int_0^{2\pi} x(s)\, ds,$$

$$a_n = \frac{1}{\pi} \int_0^{2\pi} x(s) \cos ns\, ds,$$

$$b_n = \frac{1}{\pi} \int_0^{2\pi} x(s) \sin ns\, ds,$$

$$n = 1, 2, \ldots. \qquad (6.33)$$

If y is a nonzero vector in a Hilbert space H, the *unit vector $\hat{y}$ in the direction of y* is defined as

$$\hat{y} = \frac{y}{\|y\|}, \qquad (6.34)$$

that is, the scalar $1/\|y\|$ times the vector y. Obviously $\|\hat{y}\| = 1$. The *orthogonal projection $p(x, y)$ of a vector x onto a nonzero vector y* is defined to be the vector

$$p(x, y) = \langle x, \hat{y} \rangle \hat{y} = \frac{\langle x, y \rangle}{\langle y, y \rangle} y \qquad (6.35)$$

(see Figure 6.3). The vector

$$z = x - p(x, y), \qquad (6.36)$$

as shown in Figure 6.3, is orthogonal to y. Since z is a linear combination of

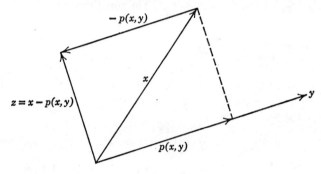

Figure 6.3

x and y by (6.34) and (6.35), with the coefficient of x being different from zero, it will not vanish if x and y are linearly independent. On the other hand, if x and y are linearly dependent, so that $x = \theta y$, then

$$p(x, y) = \theta \, \|y\| \, \hat{y} = \theta y = x, \qquad (6.37)$$

and $z = x - x = 0$. Consequently z given by (6.36) vanishes if and only if x and y are linearly dependent.

Similarly, if $\{x_1, x_2, \ldots, x_n\}$ is a set of linearly independent vectors, we can form the vectors

$$y_1 = x_1,$$

$$y_2 = x_2 - p(x_2, y_1),$$

$$y_3 = x_3 - p(x_3, y_1) - p(x_3, y_2),$$

$$\cdot$$
$$\cdot \qquad\qquad\qquad (6.38)$$
$$\cdot$$

$$y_n = x_n - \sum_{i=1}^{n-1} p(x_n, y_i),$$

which are readily seen to be mutually orthogonal and nonzero. Therefore $\{\hat{y}_1, \hat{y}_2, \ldots, \hat{y}_n\}$ is an orthonormal set. This procedure is called the *Gram-Schmidt orthonormalization process* [6]. This leads to a solution of the analysis problem in terms of an arbitrary basis $\{x_1, x_2, \ldots, x_n\}$ in the following way: The corresponding orthonormal basis $\{\hat{y}_1, \hat{y}_2, \ldots, \hat{y}_n\}$ is constructed by (6.38), which gives each $\hat{y}_i$ as a linear combination of $x_1, x_2, \ldots, x_i$,

$$\hat{y}_i = \sum_{j=1}^{i} \alpha_{ij} x_j, \qquad i = 1, 2, \ldots, n. \qquad (6.39)$$

The analysis problem is solved for

$$x = \beta_1 \hat{y}_1 + \beta_2 \hat{y}_2 + \cdots + \beta_n \hat{y}_n, \tag{6.40}$$

as indicated by (6.27); that is,

$$\beta_i = \langle x, \hat{y}_i \rangle, \qquad i = 1, 2, \ldots, n. \tag{6.41}$$

It follows from

$$x = \alpha_1 x_1 + \alpha_2 x_2 + \cdots + \alpha_n x_n \tag{6.42}$$

and (6.39) through (6.41) that

$$\alpha_j = \sum_{i=j}^{n} \langle x, \hat{y}_i \rangle \alpha_{ij}, \qquad j = 1, 2, \ldots, n, \tag{6.43}$$

which solves the problem of the analysis of x in terms of the basis

$$\{x_1, x_2, \ldots, x_n\}.$$

If the set of vectors $\{x_1, x_2, \ldots, x_n\}$ is linearly dependent, the Gram-Schmidt process will yield $y_i = 0$ for the first value of i for which x_i is a linear combination of $x_1, x_2, \ldots, x_{i-1}$, and will also give the coefficients of the combination.

If H_1 and H_2 are Hilbert spaces over the same scalar field, with inner products $\langle x_1, y_1 \rangle_1$ and $\langle x_2, y_2 \rangle_2$ respectively, it is easy to see that their product $H = H_1 \times H_2$ will be a Hilbert space for the inner product

$$\langle xy \rangle = \langle x_1, y_1 \rangle_1 + \langle x_2, y_2 \rangle_2, \tag{6.44}$$

where $x = (x_1, {}_2x)$, $y = (y_1, y_2)$ are elements of H. The definition (6.44) extends immediately to finite products of Hilbert spaces. For infinite products, conditions of convergence or integrability must be considered. As an example, if we are given Hilbert spaces H_s with the inner products $\langle x_s, y_s \rangle_s$ for $0 \le s \le 1$, we would extend (6.44) by making the usual transition from sums to integrals to define

$$\langle x, y \rangle = \int_0^1 \langle x_s, y_s \rangle_s \, ds, \tag{6.45}$$

for $x = x(s)$, $y = y(s)$ in

$$H = \prod_{0 \le s \le 1} H_s. \tag{6.46}$$

The integral in (6.45), however, will usually be meaningful and satisfy **P1**–**P3** only on a certain subspace H_s' of H_s. For H_s' to be a Hilbert space it must also be complete for the norm (6.19).

EXERCISES 6

1. Give suitable definitions of the inner product in R_2^∞, C_2^∞, and show that these spaces are Hilbert spaces.

2. Apply the Gram-Schmidt orthonormalization process to the basis x_1, x_2, x_3 for $R_2{}^3$ given in Problem 2, Exercises 3. Use the result to obtain a solution for the problem of analysis of $x = (\xi_1, \xi_2, \xi_3)$ in terms of x_1, x_2, x_3. Compare this result with the one obtained previously.

3. A sequence $\{x_m\}$ in a Hilbert space H is said to *converge weakly* to $x^* \in H$ if

$$\lim_{m \to \infty} \langle x_m, y \rangle = \langle x^*, y \rangle$$

for all $y \in H$. [The convergence defined by (4.29) is sometimes called *strong convergence*.] Show that weak and strong convergence are equivalent in $R_2{}^n$ and $C_2{}^n$. Find a sequence of unit vectors in C_2^∞ which converges weakly to 0.

4. Apply the Gram-Schmidt process to the elements

$$x_n(s) = s^n, \qquad n = 0, 1, 2, \ldots$$

of $L_2[0, 1]$ to obtain the *Legendre polynomials* $P_n(s) = \hat{y}_n(s)$, $n = 0, 1, 2, 3$ [12].

5. Compute the orthogonal projection of a function in $L_2[0, 2\pi]$ on the subspace spanned by

$$\frac{1}{\sqrt{2\pi}}, \frac{1}{\sqrt{\pi}} \sin s, \frac{1}{\sqrt{\pi}} \cos s, \ldots, \frac{1}{\sqrt{\pi}} \sin ns, \frac{1}{\sqrt{\pi}} \cos ns.$$

(The orthogonal projection of a vector on a subspace is the sum of its orthogonal projections on the vectors of a basis for the subspace.)

6. Consider the case $a = \langle z, z \rangle = 0$ in the Cauchy-Schwarz inequality (6.8). Show that equality will hold in the general case if and only if z and f are linearly dependent [8].

7. Verify that the set of functions (6.30) is an orthonormal set. [For this purpose, it is sufficient to interpret (6.31) as an ordinary Riemann integral.]

7. LINEAR OPERATORS

Many mathematical operations which transform one vector or function into another have certain simple algebraic properties in common. These properties are formulated abstractly in the following definitions.

Definitions 7.1.   An *operator L* which maps a linear space X into a linear space Y over the same scalar field Λ, so that for each $x \in X$ there is a uniquely defined element $L(x) \in Y$, is said to be *additive* if

$$L(x_1 + x_2) = L(x_1) + L(x_2), \tag{7.1}$$

for all $x_1, x_2 \in X$, and *homogeneous* if

$$L(\lambda x) = \lambda L(x), \tag{7.2}$$

for all $x \in X$, $\lambda \in \Lambda$. An operator that is additive and homogeneous is called a *linear operator*, and the notation

$$Lx = L(x) \tag{7.3}$$

will be employed for linear L. If $X = Y$, L is said to be an operator *in X*.

Many examples of linear operators may be cited. For example, scalar multiplication, with $Lx = \lambda x$ is a simple linear operator that maps a linear space X into itself. Another example of a linear operator is furnished by *differentiation*; the operator $D = d/ds$ maps $X = C'[0, 1]$ into $Y = C[0, 1]$, with

$$Dx = \frac{dx}{ds} = y(s), \qquad 0 \le s \le 1. \tag{7.4}$$

Similarly, the ordinary *differential operator* of order n, $D^n = d^n/ds^n$, is a linear operator that maps $C^{(n)}[a, b]$ into $C[a, b]$. The *Laplace operator*

$$\Delta = \frac{\partial^2}{\partial \xi^2} + \frac{\partial^2}{\partial \eta^2} + \frac{\partial^2}{\partial \zeta^2} \tag{7.5}$$

is a linear partial differential operator which maps the space of real functions $x = x(\xi, \eta, \zeta)$ with continuous second derivatives on some region in R^3 into the space of continuous real functions on that region.

The *Fourier transform F* defined by

$$Fx = \int_{-\infty}^{+\infty} e^{2\pi i s t}\, x(t)\, dt = y(s), \qquad -\infty < s < +\infty, \tag{7.6}$$

maps the space of complex-valued functions $x(s)$ on $-\infty < s < +\infty$ having certain integrability properties into itself. If x and y are interpreted not as functions but as mathematical objects called *distributions* or *generalized functions* [13], the theory of (7.6) may be cast in a very simple form.

If $X = R^n$, $Y = R^m$, then a linear operator L has a unique representation as an $m \times n$ *matrix* [2],

$$L = \begin{pmatrix} \lambda_{11} & \lambda_{12} & \cdots & \lambda_{1n} \\ \lambda_{21} & \lambda_{22} & \cdots & \lambda_{2n} \\ \cdot & & & \\ \cdot & & & \\ \cdot & & & \\ \lambda_{m1} & \lambda_{m2} & \cdots & \lambda_{mn} \end{pmatrix}. \tag{7.7}$$

Here the correspondence $Lx = y$ has the form

$$\eta_i = \sum_{j=1}^{n} \lambda_{ij}\xi_j, \qquad i = 1, 2, \ldots, m, \tag{7.8}$$

for $x = (\xi_1, \xi_2, \ldots, \xi_n)$, $y = (\eta_1, \eta_2, \ldots, \eta_m)$. Another symbolic representation of $Lx = y$ is

$$\begin{pmatrix} \lambda_{11} & \lambda_{12} & \cdots & \lambda_{1n} \\ \lambda_{21} & \lambda_{22} & \cdots & \lambda_{2n} \\ \cdot & & & \\ \cdot & & & \\ \cdot & & & \\ \lambda_{m1} & \lambda_{m2} & \cdots & \lambda_{mn} \end{pmatrix} \begin{pmatrix} \xi_1 \\ \xi_2 \\ \cdot \\ \cdot \\ \cdot \\ \xi_n \end{pmatrix} = \begin{pmatrix} \eta_1 \\ \eta_2 \\ \cdot \\ \cdot \\ \cdot \\ \eta_m \end{pmatrix}, \tag{7.9}$$

which expresses the *row-by-column rule* for operation by a matrix on a vector. For any two finite-dimensional spaces X and Y, the representation (7.7) holds once bases $\{x_1, x_2, \ldots, x_n\}$ and $\{y_1, y_2, \ldots, y_m\}$ have been chosen for X and Y respectively [2].

In function spaces the summation in (7.8) would be replaced by the operation of integration, and the matrix $L = (\lambda_{ij})$ by the *kernel* $L(s, t)$ of an *integral transform* such as

$$Lx(s) = \int_a^b L(s, t) \, x(t) \, dt = y(s), \qquad a \le s \le b, \qquad (7.10)$$

which will map $C[a, b]$ into itself if $L(s, t)$ is continuous for $a \le s, t \le b$. The Fourier transform (7.6) is also of this type, with

$$L(s, t) = e^{2\pi i s t}, \qquad -\infty < s, t < +\infty. \qquad (7.11)$$

If $X = \prod_{i=1}^{n} X_i$, the cartesian product of n linear spaces, and $Y = \prod_{i=1}^{m} Y_i$, all spaces being over a common scalar field, then a linear operator L from X into Y may be represented uniquely as the matrix array

$$L = \begin{pmatrix} L_{11} & L_{12} & \cdots & L_{1n} \\ L_{21} & L_{22} & \cdots & L_{2n} \\ \cdot & & & \\ \cdot & & & \\ L_{m1} & L_{m2} & \cdots & L_{mn} \end{pmatrix}, \qquad (7.12)$$

where L_{ij} is a linear operator from X_j into Y_i, $i = 1, 2, \ldots, m, j = 1, 2, \ldots, n$. The transformation $y = Lx$, where

$$y_i = \sum_{j=1}^{n} L_{ij} x_j, \qquad i = 1, 2, \ldots, m, \qquad (7.13)$$

has the symbolic representation

$$\begin{pmatrix} y_1 \\ y_2 \\ \cdot \\ \cdot \\ \cdot \\ y_m \end{pmatrix} = \begin{pmatrix} L_{11} & L_{12} & \cdots & L_{1n} \\ L_{21} & L_{22} & \cdots & L_{2n} \\ \cdot & & & \\ \cdot & & & \\ L_{m1} & L_{m2} & \cdots & L_{mn} \end{pmatrix} \begin{pmatrix} x_1 \\ x_2 \\ \cdot \\ \cdot \\ x_n \end{pmatrix}. \qquad (7.14)$$

For products of a countably infinite number of spaces the representation (7.12) of a linear operator L would be extended to an infinite matrix. For an uncountable number of spaces we would pass to the representation of linear operators as kernels of integral transforms, similar to (7.10). The kernel $L(s, t)$ discussed in connection with (7.10) can actually be regarded to be a limiting case of a matrix, $L(s, t)$ having a numerical *entry* (component) for each point (s, t) in the square $a \leq s, t \leq b$, just as the finite matrix (7.7) has the entry λ_{ij} for each i, j such that $1 \leq i \leq m$, $1 \leq j \leq n$.

An example of an operator of the form (7.12) is the operator

$$L = \left(a \frac{d^2}{ds^2} \quad b \frac{d}{ds} \right), \tag{7.15}$$

which maps $C''[0, 1] \times C'[0, 1]$ into $C[0, 1]$ by

$$L(x, y) = \left(a \frac{d^2}{ds^2} \quad b \frac{d}{ds} \right) \binom{x}{y} = a \frac{d^2 x}{ds^2} + b \frac{dy}{ds}, \quad 0 \leq s \leq 1, \tag{7.16}$$

where $x = x(s)$ is an element of $C''[0, 1]$, and $y = y(s)$ belongs to $C'[0, 1]$.

The representation of an element (x, y) of a product space by $\binom{x}{y}$ is said to be its *column vector* form; (x, y) is called a *row vector*.

If X and Y are linear spaces over a common scalar field Λ, then the set $L(X, Y)$ of all linear operators from X into Y becomes a linear space over Λ if addition is defined by

$$(L_1 + L_2)x = L_1 x + L_2 x \tag{7.17}$$

for all $x \in X$, and scalar multiplication by

$$(\lambda L)x = \lambda(Lx), \tag{7.18}$$

for all $x \in X$, $\lambda \in \Lambda$. The satisfaction of conditions **A1–A4**, **M1–M3** for these definitions follows immediately from the fact that Y is a linear space, so that, for example,

$$(L_1 + L_2)x = L_1 x + L_2 x = L_2 x + L_1 x = (L_2 + L_1)x, \tag{7.19}$$

for all $x \in X$, from which

$$L_1 + L_2 = L_2 + L_1, \tag{7.20}$$

thus verifying (1.3).

Since $L(X, Y)$ as previously defined is a linear space over the same scalar field as X, we may consider linear operators B which map X into $L(X, Y)$. For such an operator B and an $x \in X$ we would have

$$Bx = L, \tag{7.21}$$

a linear operator from X into Y. Thus for $x_1, x_2 \in X$, we would obtain

$$Bx_1 x_2 = (Bx_1)x_2 = y, \tag{7.22}$$

a point belonging to Y. Sometimes B is called a *bilinear* operator from X into

Y. Since the linear operators B from X into $L(X, Y)$ themselves form a linear space $L(X, L(X, Y))$, the foregoing process could be repeated, leading to a whole hierarchy of linear operators and spaces. These classes of operators play a fundamental role in the differential calculus in Banach spaces, which is discussed in a later chapter.

Definition 7.2. A linear operator which maps a linear space X into its scalar field Λ is called a *linear functional* on X.

The operator δ_0 introduced in (2.6) is a linear functional on $C'[0, T]$, and is also a linear functional on $C[0, T]$ for the same definition. Another linear functional on $C[0, T]$ is the (definite) integral J, where

$$Jx = \int_0^T x(s)\, ds, \tag{7.23}$$

for all $x \in C[0, T]$.

Definition 7.3. An operator P from a linear space X into a linear space Y is said to be *nonlinear* if it is not a linear operator from X into Y.

A simple nonlinear operator is one that gives, for all $x \in X$,

$$P(x) = y_0, \tag{7.24}$$

where y_0 is a fixed, nonzero element of Y. Many other nonlinear operators will be discussed later.

If P is an operator, not necessarily linear, that maps a linear space X into a linear space Y, then

$$R(P) = \{y: P(x) = y, x \in X\} \tag{7.25}$$

is a subset of Y which is called the *range* of P.

If $P = L$ is a linear operator, $R(L)$ is a linear subspace of Y. If $R(L) = Y$, L is said to map X *onto* Y. For a linear operator L, the set

$$N(L) = \{x: Lx = 0, x \in X\} \tag{7.26}$$

is called the *null space* of L. $N(L)$ is always nonempty [it contains $x = 0$ by (7.2)] and is a linear subspace of X. If $N(L) = \{0\}$, that is, if the null space of L consists of only the origin 0 of X, then L is said to be *nonsingular*. In this case, the equation $Lx = y$ has only one solution of a given $y \in R(L)$.

Two important linear operators that map a linear space X into itself are the *identity operator* I such that

$$Ix = x \tag{7.27}$$

for all $x \in X$ and the *zero operator* 0 such that

$$0x = 0 \tag{7.28}$$

for all $x \in X$. Obviously

$$R(I) = N(0) = X,$$
$$N(I) = R(0) = \{0\}. \tag{7.29}$$

EXERCISES 7

1. Show that, if P is an operator that is either additive or homogeneous, then $P(0) = 0$. Define an operator from R^2 into R that is homogeneous but not additive. Define an operator in the space C of complex numbers that is additive but not homogeneous.

2. Show that $L(R^2, R^2)$ is the space of real 2×2 matrices, with

$$A + B = \begin{pmatrix} a_{11} & a_{12} \\ a_{21} & a_{22} \end{pmatrix} + \begin{pmatrix} b_{11} & b_{12} \\ b_{21} & b_{22} \end{pmatrix} = \begin{pmatrix} a_{11} + b_{11} & a_{12} + b_{12} \\ a_{21} + b_{21} & a_{22} + b_{22} \end{pmatrix},$$

and

$$\lambda A = \begin{pmatrix} \lambda a_{11} & \lambda a_{12} \\ \lambda a_{21} & \lambda a_{22} \end{pmatrix}.$$

Find a basis for $L(R^2, R^2)$.

3. Show that a linear operator B from R^2 into $L(R^2, R^2)$ can be represented by the array

$$B = \begin{pmatrix} b_{111} & b_{112} & b_{121} & b_{122} \\ b_{211} & b_{212} & b_{221} & b_{222} \end{pmatrix},$$

where

$$Bx = \begin{pmatrix} b_{111} & b_{112} & b_{121} & b_{122} \\ b_{211} & b_{212} & b_{221} & b_{222} \end{pmatrix} \begin{pmatrix} \xi_1 \\ \xi_2 \end{pmatrix}$$

$$= \begin{pmatrix} b_{111}\xi_1 + b_{112}\xi_2 & b_{121}\xi_1 + b_{122}\xi_2 \\ b_{211}\xi_1 + b_{212}\xi_2 & b_{221}\xi_1 + b_{222}\xi_2 \end{pmatrix}$$

is the matrix formed by extending the row-by-column rule. Show that a linear operator B from R^n into $L(R^n, R^m)$ [or C^n into $L(C^n, C^m)$] will have the form

$$B = (b_{ijk}), \quad i = 1, 2, \ldots, m; \quad j, k = 1, 2, \ldots, n,$$

where, for $x = (\xi_1, \xi_2, \ldots, \xi_n)$, the matrix $A = Bx$ has coefficients given by

$$A = (a_{ij}) = \left(\sum_{k=1}^{n} b_{ijk}\xi_k \right), \quad i = 1, 2, \ldots, m; \quad j = 1, 2, \ldots, n.$$

4. Show that the operator $D = d/ds$ from $C'[0, 1]$ into $C[0, 1]$ is *singular* that is, it has a null space $N(D)$ containing nonzero elements. Determine $N(D)$.

5. Show that the operator

$$L = \begin{pmatrix} D \\ \delta_0 \end{pmatrix}$$

from $C'[0, 1]$ into $C[0, 1] \times R$ is nonsingular, f where $D = d/ds$. Find the unique solution $x \in C'[0, 1]$ of the equation

$$\begin{pmatrix} D \\ \delta_0 \end{pmatrix} x = \begin{pmatrix} y \\ \xi \end{pmatrix}$$

for $y \in C[0, 1]$, $\xi \in R$.

6. In an inner product space show that for any fixed ϕ in the space,

$$f(z) = \langle z, \phi \rangle$$

is a linear functional. Is

$$f(z) = \langle \phi, z \rangle$$

a linear functional for fixed ϕ?

8. BOUNDED LINEAR OPERATORS

The previous section was concerned with algebraic properties of linear operators. Certain metric notions of importance in applications will be introduced in the following first for operators in general, followed by specialization to linear operators.

Definition 8.1.   An operator P from a Banach space X into a Banach space Y is *continuous at* x^* if

$$\lim_{m \to \infty} \|x_m - x^*\| = 0 \tag{8.1}$$

implies that

$$\lim_{m \to \infty} \|P(x_m) - P(x^*)\| = 0. \tag{8.2}$$

The norm in (8.1) is the norm for the space X, and the norm in (8.2) is the norm for the space Y.

This definition is a straightforward generalization of the idea of a continuous function in ordinary analysis.

Theorem 8.1.   If a linear operator L from a Banach space X into a Banach space Y is continuous at $x^* = 0$, then it is continuous at every point x of the space X.

PROOF.   Since $L(0) = 0$ for linear L, we have by assumption that

$$\lim_{m \to \infty} \|x_m\| = 0 \quad \text{implies} \quad \lim_{m \to \infty} \|Lx_m\| = 0. \tag{8.3}$$

For any x^* in X, if $\{x_m\}$ converges to x^*, then for

$$y_m = x_m - x^*, \tag{8.4}$$

it follows that

$$\lim_{m \to \infty} \|y_m\| = 0. \tag{8.5}$$

By (8.3) this implies that

$$\lim_{m \to \infty} \|Ly_m\| = \lim_{m \to \infty} \|L(x_m - x^*)\| = \lim_{m \to \infty} \|Lx_m - Lx^*\| = 0, \tag{8.6}$$

so by Definition 8.1, L is continuous at x^*, as was to be proved.

In the proof of Theorem 8.1 it was not necessary to use the fact that the operator L is homogeneous, so the theorem could have been stated for additive operators. The following theorem, however, shows that this would not be a gain in generality in the case of real Banach spaces.

Theorem 8.2. An additive operator L from a real Banach space X into a real Banach space Y is homogeneous if it is continuous.

PROOF. Since L is additive,

$$L(px) = pL(x) \tag{8.7}$$

for any nonnegative integer p. This also holds for negative p, since

$$L(0) = L(px - px) = L(px) + L(-px) = 0. \tag{8.8}$$

For any positive integer q

$$L(px) = L\left(\frac{p}{q}x + \frac{p}{q}x + \cdots + \frac{p}{q}x\right) = qL\left(\frac{p}{q}x\right), \tag{8.9}$$

from which, in view of (8.7),

$$L\left(\frac{p}{q}x\right) = \frac{p}{q}L(x); \tag{8.10}$$

therefore it follows that L is homogeneous for rational multipliers. Suppose now that $\{p_m/q_m\}$ is a sequence of rational numbers which converges to the real number λ. Then

$$\lim_{m \to \infty} \left\| \frac{p_m}{q_m}x - \lambda x \right\| = 0, \tag{8.11}$$

and, since L is continuous,

$$\lim_{m \to \infty} \left\| \frac{p_m}{q_m}L(x) - L(\lambda x) \right\| = 0, \tag{8.12}$$

or

$$\lambda L(x) = L(\lambda x), \tag{8.13}$$

which was to be proved.

Remark 8.1. For complex X and Y, an additive operator L such that $L(ix) = iL(x)$, $i = \sqrt{-1}$, is homogeneous if it is continuous.

There are, however, continuous and homogeneous operators that are not additive (see Problem 1, Exercises 7).

Another important metric concept is the boundedness of an operator, defined here for nonlinear as well as linear operators.

Definition 8.2. An operator P from a Banach space X into a Banach space Y is *bounded on the set S* in X if there exists a constant $\mu < \infty$ such that

$$\|P(x_1) - P(x_0)\| \le \mu \|x_1 - x_0\|, \tag{8.14}$$

for all $x_0, x_1 \in S$.

The *infimum* (greatest lower bound) of the numbers μ which satisfy (8.14) for $x_0 \neq x_1$ is called the *bound* of P on S. If $S = X$, P is said to be *bounded on X*. An operator which is bounded on a ball $U(x^*, r)$ of positive radius r is obviously continuous at x^*. For linear operators the converse is also true.

Theorem 8.3 (*Banach* [9]).   A continuous linear operator L from a Banach space X into a Banach space Y is bounded on X.

PROOF.   Since L is continuous, we know that a $\delta > 0$ exists such that

$$\|L\| < 1, \tag{8.15}$$

provided that

$$\|x\| < \delta. \tag{8.16}$$

It follows for any nonzero $x \in X$ that

$$\|Lx\| \le \frac{1}{\delta} \|x\|, \tag{8.17}$$

since

$$\|\theta x\| < \delta \quad \text{for} \quad |\theta| < \frac{\delta}{\|x\|}, \tag{8.18}$$

and

$$\|L\theta x\| = |\theta| \cdot \|Lx\| < 1. \tag{8.19}$$

If we set $x = x_1 - x_0$, $\mu = 1/\delta$ in (8.17), the proof is complete.

The bound on X of a linear operator L is denoted by $\|L\|$ and is sometimes called the *norm* of L. The method used in the proof of Theorem 8.3 may be employed to establish that

$$\|L\| = \sup_{\|x\|=1} \|Lx\|. \tag{8.20}$$

Consequently, for a bounded linear operator L,

$$\|Lx\| \le \|L\| \cdot \|x\| \tag{8.21}$$

for all $x \in X$.

At this point the notation $L(X, Y)$ will be redefined to mean the set of all *bounded* linear operators from X into Y in case X and Y are both Banach spaces. It is easy to see that $L(X, Y)$ will be a linear space for the definitions (7.17) and (7.18) of addition and scalar multiplication respectively.

Theorem 8.4.   The space $L(X, Y)$ is a Banach space for the norm (8.20).

PROOF. It is evident that (8.20) defines a norm in $L(X, Y)$, so that the issue at stake is whether or not $L(X, Y)$ is complete for this norm. Suppose that $\{L_m\}$ is a Cauchy sequence in $L(X, Y)$. Then, for each $x \in X$, $\{L_m x = y_m\}$ is a Cauchy sequence in Y which has a limit $y \in Y$. Define the operator L by

$$L(x) = y \tag{8.22}$$

if y is the limit of $\{L_m x\}$. It is easy to see that the operator L defined by (8.22) is additive. To show that L is bounded, note that by adding and subtracting $L_m x_0$ and $L_m x_1$ to $L(x_1) - L(x_0)$, we obtain

$$\|L(x_1) - L(x_0)\| \le \|L(x_0) - L_m x_0\| + \|L(x_1) - L_m x_1\| + \|L_m(x_1 - x_0)\|. \tag{8.23}$$

Since $\{L_m\}$ is a Cauchy sequence, an integer m' exists such that

$$\|L_m - L_{m'}\| < 1, \tag{8.24}$$

for all $m > m'$. Consequently,

$$\|L_m\| \le \|L_m - L_{m'}\| + \|L_{m'}\| \le 1 + \|L_{m'}\| = K, \tag{8.25}$$

for all m sufficiently large. Similarly, if $x_0 \ne x_1$, then for m large enough

$$\|L(x_0) - L_m x_0\| \le K \|x_1 - x_0\|,$$
$$\|L(x_1) - L_m x_1\| \le K \|x_1 - x_0\|, \tag{8.26}$$

since $\{L_m x_0\}$ converges to $L(x_0)$ and $\{L_m x_1\}$ converges to $L(x_1)$. From (8.23), (8.25), and (8.26)

$$\|L(x_1) - L(x_0)\| \le 3K \|x_1 - x_0\|, \tag{8.27}$$

so L is bounded on X and, hence, is a bounded linear operator from X into Y, which was to be proved.

So far in this section an abstract and concise summary has been given of some important facts about bounded linear operators. Attention will now be given to some examples, with others being indicated in the exercises.

The most simple example of a bounded linear operator is furnished by multiplication by a fixed scalar λ. We can write

$$L = \lambda I, \tag{8.28}$$

where I is the identity operator defined by (7.27), and it is obvious that

$$\|L\| = |\lambda|. \tag{8.29}$$

As the next example, consider the space $X = C'[0, 1]$ with the norm

$$\|x\|' = \max_{[0,1]} |x(s)| + \max_{[0,1]} |x'(s)|, \tag{8.30}$$

where $x'(s) = dx/ds$. It is well known that $C'[0, 1]$ is a Banach space for the norm (8.30) [9]. The differential operator $D = d/ds$ maps $X = C'[0, 1]$ into

$Y = C[0, 1]$, which is a Banach space for the norm

$$\|y\| = \max_{[0,1]} |y(s)| \tag{8.31}$$

(see Problem 2, Exercises 5). In the following, unless otherwise noted, the notation previously introduced for linear spaces is used for the Banach spaces obtained from them by the introduction of a suitable norm.

From (8.30) through (8.31),

$$\|y\| = \|Dx\| = \max_{[0,1]} |x'(s)| \leq \|x\|', \tag{8.32}$$

so that D is a bounded linear operator, with

$$\|D\| \leq 1. \tag{8.33}$$

The functions

$$x_m(s) = \frac{1}{m} \sin ms, \qquad m = 1, 2, \ldots, \tag{8.34}$$

are such that

$$\|Dx_m\| = 1, \tag{8.35}$$

and

$$\|x_m\|' = 1 + \frac{1}{m}, \qquad m = 1, 2, \ldots; \tag{8.36}$$

therefore it follows from (8.33) that

$$\|D\| = 1. \tag{8.37}$$

An example of a bounded linear functional f on a Hilbert space H is given by

$$fz = \langle z, \phi \rangle, \tag{8.38}$$

for fixed $\phi \in H$. By the Cauchy-Schwarz inequality (6.8) the bound of the functional f defined by (6.38) is equal to $\|\phi\|$, the norm of the element $\phi \in H$. A famous theorem due to F. Riesz [12] states that *every* bounded linear functional on H may be represented in the form (8.38). It is customary to identify the linear functional f with the element ϕ and to use the notation

$$\phi z = \langle z, \phi \rangle. \tag{8.39}$$

EXERCISES 8

1. Show that the matrix

$$L = \begin{pmatrix} \lambda_{11} & \lambda_{12} & \cdots & \lambda_{1n} \\ \lambda_{21} & \lambda_{22} & \cdots & \lambda_{2n} \\ \cdot & & & \\ \cdot & & & \\ \cdot & & & \\ \lambda_{n1} & \lambda_{n2} & \cdots & \lambda_{nn} \end{pmatrix},$$

is a bounded linear operator in $R_p{}^n$ for $p = 1, 2, \infty$, and that for

$$p = 1, \qquad \|L\|_1 \leq \max_{(j)} \sum_{i=1}^{n} |\lambda_{ij}|,$$

$$p = 2, \qquad \|L\|_2 \leq \left(\sum_{\substack{i=1 \\ j=1}}^{n} |\lambda_{ij}|^2 \right)^{\frac{1}{2}}$$

$$p = \infty, \qquad \|L\|_\infty = \max_{(i)} \sum_{j=1}^{n} |\lambda_{ij}|.$$

2. Construct 2×2 matrices for which the first two inequalities in Problem 1 are strict.

3. Show that the matrix L defined in Problem 1 is a bounded linear operator in $R_p{}^n$ for $1 \leq p \leq \infty$ and derive an upper bound for $\|L\|_p$. *Hint.* Apply Hölder's inequality (see Problem 2, Exercises 4) to the result of (7.8).

4. If $K(s, t)$ is continuous for $0 \leq s, t \leq 1$, the *linear integral transform* $Kx = y$ defined by

$$y(s) = \int_0^1 K(s, t) x(t) \, dt, \qquad 0 \leq s \leq 1,$$

defines a bounded linear operator K from the space $C[0, 1]$ into itself, with

$$\|K\| \leq \max_{[0,1]} \int_0^1 |K(s, t)| \, dt,$$

the norm for $C[0, 1]$ being

$$\|x\| = \max_{[0,1]} |x(s)|.$$

5. Derive an *estimate* (that is, an upper bound) for $\|B\|$, where

$$B = \begin{pmatrix} b_{111} & b_{112} & b_{121} & b_{122} \\ b_{211} & b_{212} & b_{221} & b_{222} \end{pmatrix}$$

is a linear operator from R_∞^2 into $L(R_\infty^2, R_\infty^2)$, with

$$\|L\| = \max \{ |\lambda_{11}| + |\lambda_{12}|, |\lambda_{21}| + |\lambda_{22}| \}$$

for

$$L = \begin{pmatrix} \lambda_{11} & \lambda_{12} \\ \lambda_{21} & \lambda_{22} \end{pmatrix} \quad \text{in} \quad L(R_\infty^2, R_\infty^2).$$

6. Exhibit a representation of the identity operator I in R^n as a matrix. Discuss the representation of I in $C[0, 1]$ by an integral transform with kernel $\delta(s, t)$ [6, 13].

9. THE SOLUTION OF LINEAR EQUATIONS

One of the central computational problems of linear functional analysis is the solution of linear equations. This problem also arises in the study of nonlinear operator equations, since some methods of attacking nonlinear equations are based on solving an approximating linear equation or a sequence of such equations.

In a Banach space X the problem of solving a linear equation may be stated as follows: given a bounded linear operator L in X (i.e., L maps X into itself) and some element $y \in X$, we seek an $x \in X$ such that

$$Lx = y, \tag{9.1}$$

and, if any such element x exists, it will be called a *solution* of the *linear equation* (9.1).

The case of principal interest for the applications to be considered later is the one in which L is nonsingular and its range $R(L) = X$, so that (9.1) has a unique solution $x \in X$ for each given $y \in X$. Necessary and sufficient conditions for this situation to obtain as well as exact and approximate expressions for the solution x of (9.1) are derived in the following.

It is convenient to introduce the notion of the *product* (or *composition*) of operators in X. If P and Q are operators (not necessarily linear) which map X into itself, then their product PQ is the operator defined by

$$PQ(x) = P(Q(x)) \tag{9.2}$$

for all $x \in X$.

For the case $P = A = (a_{ij})$, $Q = B = (b_{ij})$, $i, j = 1, 2, \ldots, n$ are matrices representing linear operators in R^n, it follows from (7.8) that their product $C = AB$ is represented by the matrix $C = (c_{ij})$, with

$$c_{ij} = \sum_{k=1}^{n} a_{ik} b_{kj}, \qquad i, j = 1, 2, \ldots, n. \tag{9.3}$$

Similarly, if $L_1 = L_1(s, t)$ and $L_2 = L_2(s, t)$ are kernels of linear integral transforms L_1, L_2 of the form (7.10) in $C[0, 1]$, then their product $L = L_1 L_2$ has the kernel $L = L(s, t)$ given by

$$L(s, t) = \int_0^1 L_1(s, u)\, L_2(u, t)\, du, \qquad 0 \le s, t \le 1. \tag{9.4}$$

Note that generally $L_1 L_2 \ne L_2 L_1$.

If L_1 and L_2 are bounded linear operators in X, it is easy to see from (8.21) that

$$\|L_1 L_2\| \le \|L_1\| \cdot \|L_2\|. \tag{9.5}$$

Throughout the symbol I will be used to denote the *identity* operator in X, that is, $Ix = x$ for all $x \in X$. Positive integral powers of an operator P, linear or nonlinear, are *defined inductively* by $P^n = PP^{n-1}$, with $P^0 = I$. If $P = L$ is a bounded linear operator in X, it follows from (9.5) that

$$\|L^n\| \le \|L\|^n, \qquad n = 0, 1, 2, \ldots. \tag{9.6}$$

The linear scalar equation $ax = y$ is solved for a $a \ne 0$ by multiplying both sides of it by the *reciprocal* $1/a$ of a to obtain $x = (1/a)y$. This idea may be extended to bounded linear operators.

Definition 9.1. If L is a bounded linear operator in X, and a bounded linear operator L^{-1} exists such that

$$L^{-1}L = LL^{-1} = I, \qquad (9.7)$$

then L^{-1} is called the *inverse* of L.

The inverse of a bounded linear operator is thus a generalization of the notion of the reciprocal of a nonzero scalar. Nonzero linear operators, however, need not have inverses; for example, the matrix

$$A = \begin{pmatrix} 2 & -1 \\ -4 & 2 \end{pmatrix} \qquad (9.8)$$

is a bounded, nonzero linear operator in R_2^2, yet no real numbers a, b, c, d exist such that

$$\begin{pmatrix} 2 & -1 \\ -4 & 2 \end{pmatrix}\begin{pmatrix} a & b \\ c & d \end{pmatrix} = \begin{pmatrix} 1 & 0 \\ 0 & 1 \end{pmatrix} = I. \qquad (9.9)$$

It follows from (9.7) that L is the inverse of L^{-1}, provided that L^{-1} exists. If L^{-1} exists, then (9.1) has the unique solution

$$x = L^{-1}y, \qquad (9.10)$$

for each $y \in X$, and

$$\|x\| \le \|L^{-1}\| \cdot \|y\|. \qquad (9.11)$$

If (9.1) has a unique solution x for each given $y \in X$, this defines a linear operator L_1 by the correspondence $x = L_1 y$. If L_1 is bounded, then L^{-1} exists and $L^{-1} = L_1$.

The fundamental theorem giving necessary and sufficient conditions for the existence of the inverse of a bounded linear operator L in X is the following [15]. It is based on the simple observation that, if $|1 - a| < 1$, then

$$\frac{1}{a} = 1 + (1 - a) + (1 - a)^2 + \cdots = \sum_{n=0}^{\infty} (1 - a)^n. \qquad (9.12)$$

Theorem 9.1. If L is a bounded linear operator in X, L^{-1} exists if and only if there is a bounded linear operator K in X such that K^{-1} exists and

$$\|I - KL\| < 1. \qquad (9.13)$$

If L^{-1} exists, then

$$L^{-1} = \sum_{n=0}^{\infty} (I - KL)^n K \qquad (9.14)$$

and

$$\|L^{-1}\| \le \frac{\|K\|}{1 - \|I - KL\|}. \qquad (9.15)$$

PROOF.   To prove the sufficiency assume that K, K^{-1} exist and that (9.13) is satisfied. Thus, since

$$\left\| \sum_{n=0}^{\infty} (I - KL)^n K \right\| \leq \|K\| \sum_{n=0}^{\infty} \|I - KL\|^n = \frac{\|K\|}{1 - \|I - KL\|}, \quad (9.16)$$

the infinite series in (9.14) defines a bounded linear operator in X. Hence for each $y \in X$

$$x^* = \sum_{n=0}^{\infty} (I - KL)^n Ky \quad (9.17)$$

is a uniquely defined element of X. From (9.17)

$$(I - KL)x^* = \sum_{n=1}^{\infty} (I - KL)^n Ky = x^* - Ky, \quad (9.18)$$

and thus

$$KLx^* = Ky. \quad (9.19)$$

Since K^{-1} exists, $Lx^* = y$, so that the equation $Lx = y$ has at least one solution $x = x^*$ for each $y \in X$. To show that this solution is unique assume that $Lx = y$ for $x \neq x^*$. Then $L(x - x^*) = 0$ and

$$(I - KL)(x - x^*) = x - x^* \neq 0, \quad (9.20)$$

which implies that $\|I - KL\| \geq 1$, in contradiction to the assumption that (9.13) holds. Consequently L^{-1} exists and is given by (9.14) and inequality (9.15) follows immediately from (9.16).

To prove the necessity assume that L^{-1} exists. For $K = L^{-1}$, $K^{-1} = L$ exists and

$$\|I - KL\| = \|I - I\| = 0 < 1, \quad (9.21)$$

so that (9.13) is satisfied.

The expression (9.14) is called the *Neumann series* expansion of L^{-1}. It is, of course, a generalization of the *geometric series* (9.12) to bounded linear operators.

A simple application of Theorem 9.1 is to obtain the solution of the real equation

$$ax = y \quad (9.22)$$

for integers a, y without actual division. If $a \neq 0$, then an integer p exists such that

$$|1 - 10^{-p}a| < 1. \quad (9.23)$$

Thus, applying Theorem 9.1 for $L = a$, $K = 10^{-p}$ (or 2^{-p} if binary notation is being used for the numbers involved), we obtain

$$\frac{1}{a} = \sum_{n=0}^{\infty} (1 - 10^{-p}a)^n 10^{-p} \quad (9.24)$$

and

$$x = \frac{1}{a} y = \sum_{n=0}^{\infty} (1 - 10^{-p}a)^n 10^{-p}y. \tag{9.25}$$

If the value of x is required to be accurate to q decimal places, we would compute

$$x_m = \sum_{n=0}^{m} (1 - 10^{-p}a)^n 10^{-p}y \tag{9.26}$$

for the smallest value of m such that

$$|x - x_m| \leq \frac{|1 - 10^{-p}a|^{m+1} |10^{-p}y|}{1 - |1 - 10^{-p}a|} < 5(10^{-q-1}). \tag{9.27}$$

Division can be avoided in the determination of m from (9.27) by using the relationship

$$|1 - 10^{-p}a|^{m+1} |10^{-p}y| < 5(10^{-q-1})(1 - |1 - 10^{-p}a|). \tag{9.28}$$

A more significant example is the linear integral equation of *Fredholm type* and *second kind*,

$$x(s) - \int_0^1 F(s, t)\, x(t)\, dt = y(s), \qquad 0 \leq s \leq 1, \tag{9.29}$$

where

$$F(s, t) = \begin{cases} s(1 - t), & 0 \leq s \leq t \leq 1, \\ t(1 - s), & 0 \leq t \leq s \leq 1. \end{cases} \tag{9.30}$$

This equation arises in connection with the study of the differential operator $D^2 = d^2/ds^2$, together with the boundary conditions $x(0) = x(1) = 0$ [18]. Equation 9.29 is of the form (9.1), with

$$L = I - F, \tag{9.31}$$

where I is the identity operator, and F is a linear integral transform of the form (7.10) with kernel $F(s, t)$ given by (9.30). For $K = I$ in Theorem 9.1,

$$I - KL = I - (I - F) = F. \tag{9.32}$$

We have (see Problem 4, Exercises 8),

$$\|F\| \leq \max_{[0,1]} \int_0^1 |F(s, t)|\, dt = \tfrac{1}{8}. \tag{9.33}$$

Consequently, condition (9.13) is satisfied and, from (9.14),

$$L^{-1} = (I - F)^{-1} = \sum_{n=0}^{\infty} F^n. \tag{9.34}$$

We usually write

$$(I - F)^{-1} = I + G, \tag{9.35}$$

where

$$G = \sum_{n=1}^{\infty} F^n. \tag{9.36}$$

The operator G may be represented as a linear integral transform with kernel $G(s, t)$, which is called the *resolvent kernel* of $F(s, t)$ [19]. This is so because the operators F^n, $n = 1, 2, \ldots$, are linear integral transforms with kernels $F_n(s, t)$, where

$$F_1(s, t) = F(s, t),$$
$$F_{n+1}(s, t) = \int_0^1 F(s, u) F_n(u, t)\, du, \qquad n = 1, 2, \ldots, \tag{9.37}$$

as a consequence of (9.4); for example, if $0 \le s \le t \le 1$, then

$$F_2(s, t) = \int_0^s u(1 - s)\, u(1 - t)\, du + \int_s^t s(1 - u)\, u(1 - t)\, du$$

$$+ \int_t^1 s(1 - u)\, t(1 - u)\, du$$

$$= \tfrac{1}{3}s^3(1 - s)(1 - t) + \tfrac{1}{2}(t^2 - s^2)\, s(1 - t)$$

$$+ \tfrac{1}{3}(s^3 - t^3)\, s(1 - t) + \tfrac{1}{3}st(1 - t)^3$$

$$= -\tfrac{1}{6}s(1 - t)(s^2 - 2t + t^2). \tag{9.38}$$

Similarly, for $0 \le t \le s \le 1$, it is found that

$$F_2(s, t) = -\tfrac{1}{6}t(1 - s)(t^2 - 2s + s^2). \tag{9.39}$$

The direct calculation of the resolvent kernel

$$G(s, t) = F_1(s, t) + F_2(s, t) + F_3(s, t) + \cdots, \tag{9.40}$$

to any high degree of accuracy would thus be fairly cumbersome. Of course, an electronic digital computer could be programmed to carry out the manipulations indicated.

For a given $y(s)$ it is usually computationally more efficient to generate *successive approximations* $x_m(s)$ to the solution $x(s)$ of (9.29). From (9.34)

$$x = L^{-1}y = \sum_{n=0}^{\infty} F^n y. \tag{9.41}$$

For

$$x_m = \sum_{n=0}^{m} F^n y, \qquad m = 0, 1, 2, \ldots, \tag{9.42}$$

we have

$$x_{m+1} = \sum_{n=0}^{m+1} F^n y = y + Fx_m, \qquad m = 0, 1, 2, \ldots, \qquad (9.43)$$

or, in terms of functions and integral transforms,

$$x_{m+1}(s) = y(s) + \int_0^1 F(s, t)\, x_m(t)\, dt$$

$$= y(s) + \int_0^s t(1 - s)\, x_m(t)\, dt + \int_s^1 s(1 - t)\, x_m(t)\, dt,$$

$$m = 0, 1, 2, \ldots. \quad (9.44)$$

Thus, for example, for $x_0(s) = y(s) = 1, 0 \le s \le 1$,

$$x_1(s) = 1 + \tfrac{1}{2}s(1 - s),$$
$$x_2(s) = 1 + \tfrac{1}{2}s(1 - s) + \tfrac{1}{12}s(1 - s^2) - \tfrac{1}{24}s(1 - s^3), \qquad (9.45)$$

are calculated with less labor than would be required for the application of $F_1(s, t)$ and $F_2(s, t)$ to $y(s)$ if we take into account the operations involved in the construction of $F_2(s, t)$.

The *error estimate*

$$\|x_m - x\| \le \frac{\|F\|^{n+1}\,\|y\|}{1 - \|F\|} \le \frac{1}{7}\left(\frac{1}{8}\right)^n \|y\| \qquad (9.46)$$

holds by (9.33), (9.41), and (9.42). For $y(s) = 1, \le s \le 1$, (9.46) gives

$$\|x_2 - x\| = \max_{[0,1]} |x_2(s) - x(s)| \le 0.0023. \qquad (9.47)$$

Other applications of Theorem 9.1 to integral equations of the form (9.29) may be found in [15].

On the basis of Theorem 9.1 we may develop a computational *theory* of the linear equation (9.1), which embraces the following topics:

T1. *Existence and uniqueness.* If the hypotheses of Theorem 9.1 are satisfied, a solution $x = x^*$ of the equation $Lx = y$ exists and is unique in the space X.

T2. *Construction.* The iterative process

$$x_{m+1} = Ky + (I - KL)x_m, \qquad (9.48)$$

$m = 0, 1, 2, \ldots$, gives a sequence $\{x_m\}$ of *successive approximations* which converges to x^*, starting from *any* $x_0 \in X$. To see this note that it follows from (9.48) that

$$x_m = \sum_{n=0}^{m-1} (I - KL)^n Ky + (I - KL)^m x_0. \qquad (9.49)$$

By (9.13) and (9.17) the first term on the right-hand side of (9.49) converges to x^* and the second term converges to 0 as $m \to \infty$.

T3. *Error estimation.* The speed of convergence of the sequence $\{x_m\}$ to x^* is readily apparent from the *error estimate*

$$\|x_m - x^*\| \le \frac{\|I - KL\|^m}{1 - \|I - KL\|} \|Ky\| + \|I - KL\|^m \|x_0\|, \qquad (9.50)$$

which follows at once from (9.49), (9.13), and (9.17). For $x_0 = Ky$ the more simple expressions

$$x_m = \sum_{n=0}^{m} (I - KL)^n Ky \qquad (9.51)$$

and

$$\|x_m - x^*\| \le \frac{\|I - KL\|^{m+1} \|Ky\|}{1 - \|I - KL\|} \qquad (9.52)$$

hold in place of (9.49) and (9.50) respectively.

The concept of error estimation carries with it the idea of *efficiency*. Since actual computation is inherently limited to the performance of a finite number of operations, we would wish to obtain approximations of given accuracy with the least amount of effort. Error estimates such as (9.50) and (9.52) permit us to obtain an indication of computation time required before actually embarking upon the generation of the sequence (9.48).

EXERCISES 9

1. Show that the condition (9.13) may be replaced by the requirement that
$$\|(I - KL)^p\| < 1$$
for some positive integer p. *Hint.* Establish the identity

$$\frac{1}{1 - a} = (1 + a + a^2 + \cdots + a^{p-1})(1 + a^p + a^{2p} + \cdots)$$

for scalar a with $|a^p| < 1$.

2. Show that the matrix

$$A = \begin{pmatrix} a_{11} & a_{12} & \cdots & a_{1n} \\ a_{21} & a_{22} & \cdots & a_{2n} \\ \cdot & & & \\ \cdot & & & \\ \cdot & & & \\ a_{n1} & a_{n2} & \cdots & a_{nn} \end{pmatrix}$$

has an inverse if

$$|a_{ii}| > \frac{1}{2} \sum_{j=1}^{n} |a_{ij}| > 0, \qquad i = 1, 2, \ldots, n.$$

Hint. In Theorem 9.1, take $L = A$ and

$$K = \begin{pmatrix} \dfrac{1}{a_{11}} & & & \\ & \dfrac{1}{a_{22}} & & \text{O} \\ & & \ddots & \\ & \text{O} & & \dfrac{1}{a_{nn}} \end{pmatrix}$$

3. Show that the linear integral equation of the Fredholm type and second kind in $C[0, 1]$,

$$x(s) - \lambda \int_0^1 F(s, t)\, x(t)\, dt = y(s), \qquad 0 \leq s \leq 1,$$

with $F(s, t)$ continuous on $0 \leq s, t \leq 1$, has a unique solution $x(s)$ for any given $y(s)$ in $C[0, 1]$ if

$$|\lambda| < \left[\max_{[0,1]} \int_0^1 |F(s, t)|\, dt \right]^{-1}.$$

Generalize this result to equations of the form $(I - \lambda F)x = y$ in a Banach space X, where F is a bounded linear operator on X.

4. Develop a general formula for

$$\int_0^1 F(s, t) t^n\, dt, \qquad 0 \leq s \leq 1, \qquad n = 0, 1, 2, \dots,$$

for $F(s, t)$ as given by (9.30). Use this result to obtain the successive approximations $x_1(s)$ and $x_2(S)$ to the solution $x(s)$ of (9.29) for $x_0(s) = y(s) = s^2, 0 \leq s \leq 1$.

5. Verify (9.38) and (9.39). Obtain $x_2(s)$ as given by (9.45) directly from

$$x_2(s) = 1 + \int_0^1 F_1(s, t)\, dt + \int_0^1 F_2(s, t)\, dt,$$

with $F_1(s, t) = F(s, t)$ given by (9.30), and $F_2(s, t)$ defined by (9.38) and (9.39).

6. For matrices $A = (a_{ij})$, $B = (b_{ij})$, $i, j = 1, 2, \dots, n$, let

$$a_i = (a_{i1}, a_{i2}, \dots, a_{in})$$

denote the *row vectors* of A and

$$b_j = (b_{1j}, b_{2j}, \dots, b_{nj})$$

the *column vectors* of B. Show that the element c_{ij} of the product matrix $C = AB$ is given by

$$c_{ij} = \langle a_i, b_j \rangle, \qquad i, j = 1, 2, \dots, n.$$

State the condition (9.7) that $B = A^{-1}$ in this notation.

10. INVERSION OF LINEAR OPERATORS

The basic result presented in the previous section, Theorem 9.1, led to a computational theory of linear equations in a Banach space by giving necessary and sufficient conditions for the existence of the inverse of a linear operator. In this section three theorems will be given which are equivalent to Theorem 9.1, together with examples of their applications to various computational problems.

To use Theorem 9.1 in an actual case, a suitable operator K must be found, which, by (9.13), is an approximate inverse of L in the sense that KL is close to the identity operator I. One way to obtain such an approximate inverse is to use the inverse of an operator sufficiently close to L, if known. This idea is made precise in the next theorem.

Theorem 10.1. If L is a bounded linear operator in X, L^{-1} exists if and only if there is a bounded linear operator M in X such that M^{-1} exists, and

$$\|M - L\| < \frac{1}{\|M^{-1}\|} . \tag{10.1}$$

If L^{-1} exists, then

$$L^{-1} = \sum_{n=0}^{\infty} (I - M^{-1}L)^n M^{-1} \tag{10.2}$$

and

$$\|L^{-1}\| \leq \frac{\|M^{-1}\|}{1 - \|I - M^{-1}L\|} \leq \frac{\|M^{-1}\|}{1 - \|M^{-1}\| \, \|M - L\|} . \tag{10.3}$$

PROOF.   The sufficiency is proved by taking $K = M^{-1}$ in Theorem 9.1 and noting that by (10.1),

$$\|I - M^{-1}L\| = \|M^{-1}(M - L)\| \leq \|M^{-1}\| \cdot \|M - L\| < 1, \quad (10.4)$$

so that (9.13) is satisfied. The estimate (10.3) follows from (9.15) and (10.4). The necessity is proved by taking $M = L$ if L^{-1} exists.

Banach [9] originally proved the sufficiency of (10.1) for the existence of L^{-1}, provided M^{-1} exists.

A simple example of an application of Theorem 10.1 is given for the operator

$$L = \begin{pmatrix} 5 & -1 \\ 2 & 4 \end{pmatrix} \tag{10.5}$$

in R^2_∞. For

$$M = \begin{pmatrix} 5 & 0 \\ 0 & 4 \end{pmatrix}, \qquad M^{-1} = \begin{pmatrix} \frac{1}{5} & 0 \\ 0 & \frac{1}{4} \end{pmatrix}, \tag{10.6}$$

and

$$\|M - L\| = 2 < \frac{1}{\|M^{-1}\|} = 4, \tag{10.7}$$

so L^{-1} exists by Theorem 10.1 and is given by (10.2). We have

$$I - M^{-1}L = \begin{pmatrix} 0 & \frac{1}{5} \\ -\frac{1}{2} & 0 \end{pmatrix} \tag{10.8}$$

and

$$(I - M^{-1}L)^2 = \begin{pmatrix} -\frac{1}{10} & 0 \\ 0 & -\frac{1}{10} \end{pmatrix}. \tag{10.9}$$

Consequently

$$(I - M^{-1}L)^{2n} = \begin{pmatrix} (-\frac{1}{10})^n & 0 \\ 0 & (-\frac{1}{10})^n \end{pmatrix} \tag{10.10}$$

and

$$(I - M^{-1}L)^{2n+1} = \begin{pmatrix} 0 & \frac{1}{5}(-\frac{1}{10})^n \\ -\frac{1}{2}(-\frac{1}{10})^n & 0 \end{pmatrix}, \qquad n = 0, 1, 2, \ldots. \tag{10.11}$$

Thus

$$\sum_{n=0}^{\infty} (I - M^{-1}L)^n = \sum_{n=0}^{\infty} \begin{pmatrix} (-\frac{1}{10})^n & \frac{1}{5}(-\frac{1}{10})^n \\ -\frac{1}{2}(-\frac{1}{10})^n & (-\frac{1}{10})^n \end{pmatrix}$$

$$= \begin{pmatrix} \frac{10}{11} & \frac{2}{11} \\ -\frac{5}{11} & \frac{10}{11} \end{pmatrix} \tag{10.12}$$

and

$$L^{-1} = \sum_{n=0}^{\infty} (I - M^{-1}L)^n M^{-1} = \begin{pmatrix} \frac{10}{11} & \frac{2}{11} \\ -\frac{5}{11} & \frac{10}{11} \end{pmatrix} \begin{pmatrix} \frac{1}{5} & 0 \\ 0 & \frac{1}{4} \end{pmatrix} = \begin{pmatrix} \frac{2}{11} & \frac{1}{22} \\ -\frac{1}{11} & \frac{5}{22} \end{pmatrix}. \tag{10.13}$$

This result may be verified directly by the use of (9.7). See also Problem 2, Exercises 9.

Another theorem that is equivalent to Theorem 9.1 is the following.

Theorem 10.2. A bounded linear operator L in a Banach space X has an inverse L^{-1} if and only if bounded linear operators K, K^{-1} exist such that the series

$$\sum_{n=0}^{\infty} (I - KL)^n K \tag{10.14}$$

converges, in which case its sum is L^{-1}.

PROOF. By the same reasoning as used in the proof of Theorem 9.1, if the series (10.14) is convergent, then it converges to L^{-1}. The necessity of the existence of K, K^{-1}, and the convergence of (10.14) is established, as in the proof of Theorem 9.1, by taking $K = L^{-1}$, when it exists.

To see the equivalence of Theorem 9.1 and Theorem 10.2 more clearly, note that (9.13) implies the convergence of the series (10.14). On the other hand, the convergence of (10.14) provides L^{-1} and (9.13) will be satisfied for $K = L^{-1}$.

At first glance Theorem 10.2 may appear to be more general than Theorem 9.1, in spite of their equivalence, since the series (10.14) may converge under less restrictive assumptions than (9.13) [16]. From a computational standpoint, however, some additional knowledge concerning the speed of convergence of the series (10.14) is needed before an error estimate such as (9.50) becomes available. Condition (9.13) also has the advantage of being subject to a direct computational check, which may not be true of the convergence of the series (10.14).

To illustrate an application of Theorem 10.2, consider the linear integral equation of the *Volterra type* and *second kind*,

$$x(t) = \int_0^t V(t, s)\, x(s)\, ds + y(t), \qquad t \geq 0 \tag{10.15}$$

where $y \in C[0, t]$ and

$$|V(t, s)| \leq m \tag{10.16}$$

for $0 \leq s \leq t$. The integral transform in (10.15) defines a linear operator V which maps $C[0, t]$ into itself, so that (10.15) may be written in the form $Lx = y$, where

$$L = I - V, \tag{10.17}$$

a linear operator in $C[0, t]$. Choosing $K = I$,

$$I - KL = V \tag{10.18}$$

and

$$\|V\| = \max_{[0,t]} \int_0^t |V(t, s)|\, ds \leq mt, \tag{10.19}$$

so that for $t < 1/m$, $\|I - KL\| = \|V\| < 1$ and Theorem 9.1 may be used directly. However, suppose that t is arbitrary. Then note that

$$\|V^2\| \leq \max_{[0,t]} \int_0^t \left| \int_0^s V(t, u)\, V(u, s)\, du \right| ds \leq \frac{m^2 t^2}{2}. \tag{10.20}$$

If we assume that

$$\|V^n\| \leq \frac{m^n t^n}{n!}, \tag{10.21}$$

it follows in the same way that

$$\|V^{n+1}\| \leq \max_{[0,t]} \int_0^t m \frac{m^n s^n}{n!}\, ds \leq \frac{m^{n+1} t^{n+1}}{(n+1)!}, \tag{10.22}$$

so that (10.21) is true for all positive integers n by mathematical induction. Consequently

$$\sum_{n=0}^{\infty} \|V^n\| \leq \sum_{n=0}^{\infty} \frac{(mt)^n}{n!} = e^{mt}, \tag{10.23}$$

which implies that

$$\sum_{n=0}^{\infty} V^n = L^{-1} \tag{10.24}$$

converges, giving the solution $x(t)$ of (10.15) for any $t \geq 0$ as the limit of the sequence

$$x_{n+1}(t) = \int_0^t V(t,s)\, x_n(s)\, ds + y(t), \qquad m = 0, 1, 2, \ldots, \tag{10.25}$$

with $x_0(t) = y(t)$. From (10.23) we obtain the error bound

$$\|x_n(t) - x(t)\| \leq \left(e^{mt} - \sum_{k=0}^{n} \frac{(mt)^k}{k!} \right) \|y\|, \tag{10.26}$$

where

$$\|y\| = \max_{[0,t]} |y(s)|. \tag{10.27}$$

It is possible to state Theorem 9.1 in still another form which is sometimes useful in applications.

Definition 10.1. A linear operator N in a Banach space X is said to be *nilpotent* if

$$N^k = 0 \tag{10.28}$$

for some positive integer k.

For example, the 2×2 matrix

$$N = \begin{pmatrix} 0 & 0 \\ 100 & 0 \end{pmatrix} \tag{10.29}$$

is nilpotent, as $N^2 = 0$.

Theorem 10.3. A bounded linear operator L in a Banach space X has an inverse L^{-1} if and only if bounded linear operators K, K^{-1} exist such that $I - KL$ is nilpotent.

PROOF. If K, K^{-1} exist and $I - KL$ is nilpotent, the series

$$\sum_{n=0}^{\infty} (I - KL)^n K = \sum_{n=0}^{k-1} (I - KL)^n K \tag{10.30}$$

converges and its sum is L^{-1} by Theorem 10.2. If L^{-1} exists, then for $K = L^{-1}$, $K^{-1} = L$ exists, and $I - KL = I - L^{-1}L = 0$ is nilpotent.

Corollary 10.1. If

$$L = I - N, \tag{10.31}$$

N nilpotent, then L^{-1} exists.

PROOF. Take $K = I$ in Theorem 10.3.

The importance of Theorem 10.3 is that it leads to the possibility of representing L^{-1} exactly with a finite number of operations; for example, with L given by (10.5), we could take

$$K = \begin{pmatrix} \frac{1}{5} & 0 \\ -\frac{1}{11} & \frac{5}{22} \end{pmatrix}, \tag{10.32}$$

giving

$$I - KL = \begin{pmatrix} 1 & 0 \\ 0 & 1 \end{pmatrix} - \begin{pmatrix} 1 & -\frac{1}{5} \\ 0 & 1 \end{pmatrix} = \begin{pmatrix} 0 & \frac{1}{5} \\ 0 & 0 \end{pmatrix}, \tag{10.33}$$

a nilpotent matrix. Thus

$$L^{-1} = \sum_{n=0}^{1} (I - KL)^n K = \begin{pmatrix} 1 & \frac{1}{5} \\ 0 & 1 \end{pmatrix} \begin{pmatrix} \frac{1}{5} & 0 \\ -\frac{1}{11} & \frac{5}{22} \end{pmatrix} = \begin{pmatrix} \frac{2}{11} & \frac{1}{22} \\ -\frac{1}{11} & \frac{5}{22} \end{pmatrix}, \tag{10.34}$$

as found before. It is natural to inquire how the matrix K given by (10.32) was found. In the case of finite matrices, we can employ an efficient computational method known as *gaussian elimination* [17] to arrive at such matrices. For example, consider the 2×2 matrix,

$$A = \begin{pmatrix} a_{11} & a_{12} \\ a_{21} & a_{22} \end{pmatrix}. \tag{10.35}$$

If $a_{11} \neq 0$, the first row of A is divided by a_{11}, which yields

$$A_1 = K_1 A = \begin{pmatrix} \dfrac{1}{a_{11}} & 0 \\ 0 & 1 \end{pmatrix} \begin{pmatrix} a_{11} & a_{12} \\ a_{21} & a_{22} \end{pmatrix} = \begin{pmatrix} 1 & \dfrac{a_{12}}{a_{11}} \\ a_{21} & a_{22} \end{pmatrix}. \tag{10.36}$$

(Of course, if $a_{11} = 0$, we would divide by a_{12} to obtain a similar result. If $a_{11} = a_{12} = 0$, then A is singular.) We then subtract a_{21} times the first row of A_1 from its second row to obtain

$$A_2 = K_2 A_1 = K_2 K_1 A = \begin{pmatrix} 1 & 0 \\ -a_{21} & 1 \end{pmatrix} \begin{pmatrix} 1 & \dfrac{a_{12}}{a_{11}} \\ a_{21} & a_{22} \end{pmatrix}$$

$$= \begin{pmatrix} 1 & \dfrac{a_{12}}{a_{11}} \\ 0 & a_{22} - \dfrac{a_{12}a_{21}}{a_{11}} \end{pmatrix}. \tag{10.37}$$

If

$$d = a_{11}a_{22} - a_{12}a_{21} \neq 0, \tag{10.38}$$

we divide the second row of A_2 by

$$\frac{a_{11}a_{22} - a_{12}a_{21}}{a_{11}} = \frac{d}{a_{11}}, \tag{10.39}$$

so that

$$A_3 = K_3 A_2 = K_3 K_2 K_1 A = \begin{pmatrix} 1 & 0 \\ 0 & \dfrac{a_{11}}{d} \end{pmatrix} \begin{pmatrix} 1 & \dfrac{a_{12}}{a_{11}} \\ 0 & \dfrac{d}{a_{11}} \end{pmatrix} = \begin{pmatrix} 1 & \dfrac{a_{12}}{a_{11}} \\ 0 & 1 \end{pmatrix}. \tag{10.40}$$

The matrix A_3 is of the form $I - N$, N nilpotent, and

$$A_3 = KA, \tag{10.41}$$

where

$$K = K_3 K_2 K_1 = \begin{pmatrix} 1 & 0 \\ 0 & \dfrac{a_{11}}{d} \end{pmatrix} \begin{pmatrix} 1 & 0 \\ -a_{21} & 1 \end{pmatrix} \begin{pmatrix} \dfrac{1}{a_{11}} & 0 \\ 0 & 1 \end{pmatrix}$$

$$= \begin{pmatrix} \dfrac{1}{a_{11}} & 0 \\ -\dfrac{a_{21}}{d} & \dfrac{a_{11}}{d} \end{pmatrix}. \tag{10.42}$$

For this value of K, A^{-1} can be expressed in a fashion similar to (10.34).

EXERCISES 10

1. Assuming that L_0^{-1} exists, show that the operator $L = L_0 + \lambda M$ has an inverse for

$$|\lambda| < \frac{1}{\|L_0^{-1}\| \cdot \|M\|}.$$

Write the Neumann series for L^{-1}.

2. Establish the identity due to Euler,

$$\frac{1}{1-a} = 1 + a + a^2 + \cdots = \prod_{n=0}^{\infty} (1 + a^{2^n}) = (1 + a)(1 + a^2)(1 + a^4)\cdots,$$

for scalar a with $|a| < 1$. Extend this to the Neumann series for L^{-1}, assuming that (9.13) is satisfied. Compare the rate of convergence of the partial sums of the Neumann series with the rate of convergence of the corresponding partial products.

3. Show that the matrix K given by (10.32) may be expressed in the form (10.42), where A is the matrix (10.5).

4. Prove that, if L is a bounded linear operator from a Banach space X into a Banach space Y, and there exists a nonsingular bounded linear operator K from Y onto X, that is, the range of K is X, such that $\|I - KL\| < 1$, where I is the identity operator in X, then the equation

$$Lx = y$$

has the unique solution

$$x = \sum_{n=0}^{\infty} (I - KL)^n Ky$$

for each $y \in Y$. Consequently L will be nonsingular and its range will be Y. Note that $I - KL$ is an operator in X.

5. Use the result of Problem 4 to "invert" the linear differential operator $\lambda I - D$, $D = d/ds$, which maps $X = C'[0, 1]$ into $Y = C[0, 1]$, for $|\lambda| < 1$. *Hint.* The operator defined by

$$x(s) = y(0) + \int_0^s y(t)\,dt, \qquad 0 \le s \le 1,.$$

is a nonsingular mapping from Y onto X.

6. Show that the gaussian elimination process will lead to an inverse of the matrix $A = (a_{ij})$ of order n if and only if the row vectors $a_i = (a_{i1}, a_{i2}, \ldots, a_{in})$, $i = 1, 2 \ldots, n$ of A are linearly independent. *Hint.* At the stages in the elimination process calling for division it may be necessary to interchange the remaining columns to obtain a nonzero divisor. This will require a factor in the expression for K different in form from the ones previously discussed.

7. In the space R_∞^∞ of bounded sequences the *shift operator* E is defined by

$$Ex = (0, \xi_1, \xi_2, \xi_3, \ldots)$$

for $x = (\xi_1, \xi_2, \xi_3, \ldots)$. Discuss the inversion of the *difference operator*

$$\Delta = E - I \quad \text{on} \quad R_\infty^\infty.$$

8. Show that the sequence (9.48) for $x_0 = Ky$ is also defined for $x_{-1} = 0$, $x_0 = Ky$, by $x_{m+1} = x_m + (I - KL)(x_m - x_{m-1})$. Apply this to (10.25).

REFERENCES AND NOTES

[1] N. Dunford and J. T. Schwartz, *Linear Operators*. Part 1: General Theory. 1958. Part 2: Self-Adjoint Operators in Hilbert Space. 1963. Interscience, New York. These two volumes, which contain almost 2000 pages of text and weigh approximately $6\frac{1}{3}$ lb, indicate the extent to which linear functional analysis has developed. A third volume is in preparation.

[2] A. E. Taylor. *Introduction to Functional Analysis*. Wiley, New York, 1958. This book is a readable introduction to linear functional analysis. The author prefers to indicate further abstractions to the reader, rather than applications.

[3] F. Riesz and B. von Sz.–Nagy. *Functional Analysis*, Ungar, New York, 1955.

[4] A. N. Kolmogorov and S. V. Fomin. *Functional Analysis*. Vol. 1: Metric and Normed Spaces. 1957. Vol. 2: Measure. The Lebesgue integral. Hilbert space. 1961. Graylock Press, Albany, New York. These little books give strong emphasis to integration theory and to the aspects of functional analysis that are most closely related to classical real-variable theory.

[5]   V. Volterra. *Theory of Functionals and of Integral and Integrodifferential Equations.* Dover, New York, 1959. This is a paperback reprint of a book which was written at a time when functional analysis was in a formative stage.

[6]   B. Friedman. *Principles and Techniques of Applied Mathematics.* Wiley, New York, 1958.

[7]   T. N. E. Greville. Spline Functions, Interpolation, and Numerical Quadrature. Chapter 8, in A. Ralston and H. S. Wilf (eds.), *Mathematical Methods for Digital Computers.* Vol. 2. Wiley, New York, 1967.

[8]   G. H. Hardy, J. E. Littlewood, and G. Pólya. *Inequalities.* Cambridge University Press, London, 1934.

[9]   S. Banach. *Théorie des opérations linéaires.* Second ed. Chelsea, New York, 1963.

[10]   D. Hilbert. *Grundzüge einer allgemeinen Theorie der linearen Integralgleichungen.* Chelsea, New York, 1953.

[11]   E. Hellinger and O. Toeplitz. *Integralgleichungen und Gleichungen mit unendlichvielen Unbekannten.* Chelsea, New York, 1953.

[12]   N. I. Akheizer and I. M. Glazman. *Theory of Linear Operators in Hilbert Space.* Vol. 1. Tr. by Merlynd Nestell. Ungar, New York, 1961.

[13]   M. J. Lighthill. *Introduction to Fourier Analysis and Generalized Functions.* Cambridge University Press, London, 1960.

[14]   L. V. Kantorovič. Functional Analysis and Applied Mathematics. *Uspehi Mat. Nauk*, 3 (1948), 89–185 (Russian). A translation of this paper into English was prepared by C. D. Benster and issued as *Natl. Bur. Std. Rept. No. 1509*, Washington, 1952. Unfortunately this report is no longer available from the Government Printing Office.

[15]   L. B. Rall. Error Bounds for Iterative Solutions of Fredholm Integral Equations. *Pacific J. Math.*, 5 (1955), 977–986.

[16]   W. V. Petryshyn. On a General Iterative Method for the Approximate Solution of Linear Operator Equations. *Math. Comp.*, 17 (1963), 1–10.

[17]   V. N. Faddeeva. *Computational Methods of Linear Algebra.* Tr. by C. D. Benster. Dover, New York, 1959.

[18]   R. Courant and D. Hilbert. *Methods of Mathematical Physics.* Vol. 1. Interscience, New York, 1953. The original German edition of this book was probably the most influential work on the applications of mathematical analysis ever published. The volume listed belongs in the library of every serious student of the subject.

[19]   W. V. Lovitt. *Linear Integral Equations.* Dover, New York, 1950.

2

THE CONTRACTION
MAPPING PRINCIPLE

This chapter is devoted to the discussion of a simple but powerful method of attacking nonlinear (and also linear) equations. This method is sometimes called the *method of successive substitutions* or the *method of iteration*. This procedure originated in antiquity, appearing, for example, in the writings of Heron of Alexandria [1] in the second century B.C. in connection with the extraction of roots. In modern times Cauchy and Picard [2] used this technique to establish the existence of solutions of differential equations. An abstract formulation of these results as the *contraction mapping principle* was achieved by Banach [3] and further elaborated by Caccioppoli [4] and Weissinger [5]. A survey of the applications of this principle to theoretical and computational problems would require volumes.

11. FIXED POINTS OF OPERATORS

Definition 11.1. If F is an operator that maps the Banach space X into itself, then any $x \in X$ such that

$$x = F(x) \tag{11.1}$$

is called a *fixed point* of the operator F.

For example, the operator $F(x) = x^2$ in the space R of real numbers has the fixed points $x = 0$ and $x = 1$. The linear operator

$$Fx = x(0) + \int_0^s x(t)\, dt \tag{11.2}$$

in $C[0, 1]$ has any function $x = x(s)$ of the form

$$x(s) = ce^s, \qquad 0 \le s \le 1, \tag{11.3}$$

as a fixed point, where c is a real constant.

Finding a fixed point of an operator F is equivalent to obtaining a solution of (11.1). The form of (11.1) suggests a technique for this purpose. If we know a value x_0 of x such that $F(x_0)$ does not differ greatly from x_0, it is natural to regard

$$x_1 = F(x_0) \tag{11.4}$$

as a possible improvement over x_0, and to generate the sequence $\{x_m\}$ of *successive approximations* to a fixed point x of F by the relationship

$$x_{m+1} = F(x_m), \qquad m = 0, 1, 2, \ldots. \tag{11.5}$$

To carry out the generation of the sequence $\{x_m\}$ by (11.5) all that is necessary in the way of computational facility is the ability to evaluate $F(x)$ for given values of x. This is the reason that the use of (11.5) is frequently called the *method of successive substitutions* or the *method of iteration*. Examples of the application of this technique to the Fredholm integral equation (9.29), and the Volterra integral equation (10.15) were given in the previous chapter.

It is usually possible to reformulate a given equation as a fixed point problem of the type (11.1); for example, if P is a given operator that maps a Banach space X into itself and $y \in X$ is given, then we may wish to find an $x \in X$ such that

$$P(x) = y. \tag{11.6}$$

More concisely, suppose that we want to *solve* (11.6) for x. For the operator

$$F(x) = x + P(x) - y \tag{11.7}$$

it is evident that (11.1) is satisfied by x if and only if x is a solution of (11.6). Thus we may approach the problem of solving an equation of the form $P(x) = y$ by transforming it into a fixed point problem and then using the iteration method (11.5) to generate successive approximations to the solution sought.

Of course, there are many ways to obtain (11.1) from (11.6) in addition to (11.7); for example, if Q is any operator in X such that

$$Q(x) = 0 \quad \text{if and only if } x = 0, \tag{11.8}$$

then we could define

$$F(x) = x + Q(P(x) - y) \tag{11.9}$$

and obtain a fixed point problem (11.1) equivalent to (11.6).

The method of formulating a fixed point problem may determine whether or not the generation of the sequence $\{x_m\}$ by (11.5) leads to anything worthwhile; for example, the equation

$$x^2 = 3 \qquad (11.10)$$

in $X = R$ may be reformulated as

$$x = \frac{3}{x}, \qquad (11.11)$$

which is of the form (11.1). If we take $x_0 = 1$, then (11.5) becomes

$$x_0 = 1, \qquad x_{m+1} = \frac{3}{x_m}, \qquad m = 0, 1, 2, \ldots, \qquad (11.12)$$

which yields the useless sequence

$$x_0 = 1, \qquad x_1 = 3, \ldots, \qquad x_{2m} = 1, \qquad x_{2m+1} = 3, \ldots, \qquad (11.13)$$

where $m = 0, 1, 2, \ldots$. This is in spite of the fact that (11.11) is satisfied by $x = \pm\sqrt{3}$, and $x_0 = 1$ is not a terribly bad approximation to $\sqrt{3}$. On the other hand, $x = \sqrt{3}$ also satisfies

$$x = \frac{1}{2}\left(x + \frac{3}{x}\right), \qquad (11.14)$$

and the corresponding sequence $\{x_m\}$ [see (5.1)] converges rapidly to $\sqrt{3}$ for $x_0 = 1$. This is the method attributed to Heron of Alexandria.

The discussion of sufficient conditions for the convergence of the sequence generated by (11.5) to a solution of (11.1) is deferred to the next section. In general, these conditions depend on both the operator F and the choice of the initial point x_0. It is also shown later that if all that is desired is an approximation to the fixed point it is permissible to evaluate $F(x_m)$ approximately in the generation of a finite number of terms of the sequence (11.5), as long as certain conditions of accuracy are met. This leads to the following important idea.

Definition 11.2. In R^n (or C^n), (11.1) is said to be an *arithmetic fixed point problem* if $F(x)$ can be evaluated, at least to any desired accuracy, by a finite number of additions, subtractions, multiplications, and divisions.

An example of this type of problem is (11.14), for which the sequence defined by (5.1) may be used to calculate $\sqrt{3}$ to any desired degree of accuracy.

The importance of arithmetic fixed point problems stems from the fact that the labor involved in generating successive terms in the sequence $\{x_m\}$ defined by (11.5) can be performed by an electronic digital computer. Thus an equation which can be expressed as an arithmetic fixed point problem or approximated with sufficient accuracy can be solved numerically, once the convergence of the sequence (11.5) and appropriate error bounds have been established.

Some operators have no fixed points, others have a positive finite number, whereas still others have an infinite number of fixed points. As an example of the last class of operators, the identity operator I has every point of the space X as a fixed point. Another example of an operator with many fixed points is a *projection operator*, which is a nonzero bounded linear operator Π in X such that

$$\Pi^2 = \Pi. \tag{11.15}$$

It follows immediately from (11.15) that every point in the range $R(\Pi)$ of Π is a fixed point of Π. It is also easy to see that

$$\|\Pi\| \geq 1. \tag{11.16}$$

If the range $R(\Pi)$ of a projection operator Π is a finite-dimensional subspace of X, then $R(\Pi)$ is equivalent to R^n (or C^n) once a basis has been chosen for it. Thus one method of approximating (11.1) would be to use the equation

$$y = \Pi F(y), \qquad y \in R(\Pi), \tag{11.17}$$

expressed in terms of a suitable basis $y_1, y_2, \ldots, y_n$ for $R(\Pi)$. The equation in R_n (or C^n) obtained in this way is called the *projection of* (11.1) *into* R^n (or C^n). In this way it is frequently possible to obtain an arithmetic fixed point problem from an *analytic* fixed point problem, that is, one which does not satisfy Definition 11.2.

To illustrate this idea consider the partial differential equation

$$\Delta u = u^2 \tag{11.18}$$

where

$$\Delta = \frac{\partial^2}{\partial \xi^2} + \frac{\partial^2}{\partial \eta^2} \tag{11.19}$$

is the two-dimensional Laplace operator. Equation (11.18) occurs in the theory of gas dynamics [6]. Suppose that (11.18) is satisfied in the interior of the square $0 \leq \xi, \eta \leq 1$ in R^2, and that $u(\xi, \eta) > 0$ is given and is continuous on the boundary of the square; for example,

$$\begin{aligned} u(\xi, 0) &= 2\xi^2 - \xi + 1, & 0 \leq \xi \leq 1, \\ u(1, \eta) &= 2, & 0 \leq \eta \leq 1, \\ u(\xi, 1) &= 2, & 0 \leq \xi \leq 1, \\ u(0, \eta) &= 2\eta^2 - \eta + 1, & 0 \leq \eta \leq 1. \end{aligned} \tag{11.20}$$

A solution of (11.18) is sought in $C^2(S)$, the space of real functions which are continuously twice differentiable in the interior of the square

$$S = \{(\xi, \eta): 0 \leq \xi, \eta \leq 1\}, \tag{11.21}$$

and are continuous on its boundary.

By analogy with (11.14), (11.18) may be formulated as the fixed point problem

$$u = (1 - \theta)u + \frac{\theta \Delta u}{u},$$ (11.22)

where θ is some real number in the interval $0 \leq \theta \leq 1$. Equation (11.14) corresponds to the case that $\theta = \frac{1}{2}$. To project this problem into a finite-dimensional space the method of discretization (or *finite differences* or *nets*) is used. For some positive integer n set

$$h = \frac{1}{n},$$ (11.23)

and replace the square $0 \leq \xi, \eta \leq 1$ by the *grid* of points

$$p_{ij} = (ih, jh), \qquad i, j = 0, 1, \ldots, n.$$ (11.24)

There are $(n + 1)^2$ of these *grid points*. so that a function defined on this grid can be considered as an element of $R^{(n+1)^2}$. A projection Π of $C(S)$ into $R^{(n+1)^2}$ can be defined by

$$u_{ij} = \Pi u(p_{ij}) = u(ih, jh), \qquad i, j = 0, 1, \ldots, n,$$ (11.25)

and

$$\Delta_h u_{ij} = \Pi \Delta u(p_{ij})$$

$$= \frac{u((i+1)h, jh) + u((i-1)h, jh) + u(ih, (j+1)h) + u(ih, (j-1)h)}{h^2}$$
$$- \frac{4u(ih, jh)}{h^2}$$
$$= \frac{1}{h^2} (u_{i+1,j} + u_{i-1,j} + u_{i,j+1} + u_{i,j-1} - 4u_{ij}),$$ (11.26)

where $i, j = 1, 2, \ldots, n - 1$ [6]. Thus (11.22) becomes

$$u_{ij} = (1 - \theta)u_{ij} + \frac{\theta \Delta_h u_{ij}}{u_{ij}}$$ (11.27)

or, for $\theta = h^2$,

$$u_{ij} = (1 - h^2)u_{ij} + \frac{u_{i+1,j} + u_{i-1,j} + u_{i,j+1} + u_{i,j-1}}{u_{ij}} - 4,$$
$$i, j = 1, 2, \ldots, n - 1.$$ (11.28)

For the simple case $n = 2$, it follows from (11.20) that

$$u_{21} + u_{01} + u_{12} + u_{10} = 6,$$ (11.29)

and (11.28) becomes

$$u_{11} = \tfrac{3}{4}u_{11} + \frac{6}{u_{11}} - 4. \tag{11.30}$$

It may be computed directly from (11.30) that

$$u_{11} = -8 \pm 2\sqrt{22}$$

$$= 1.380832, \; -17.380823, \tag{11.31}$$

to six decimal places. For the physical application that gave rise to (11.18), the negative solution of (11.30) would be rejected and

$$u_{11} = 1.380832 \tag{11.32}$$

would be taken to be an approximation to $u(\tfrac{1}{2}, \tfrac{1}{2})$. For this procedure to be useful, of course, something should be known about the error of this approximation or at least its convergence to the value of a solution of (11.18) at $(\tfrac{1}{2}, \tfrac{1}{2})$ as $h \to 0$. This discussion has been for the purpose of showing how an arithmetic fixed point problem may be obtained from an analytic one. Detailed numerical treatments of (11.18) may be found in the literature [6,7,8].

EXERCISES 11

1. Construct arithmetic fixed point problems for each of the following numbers: (a) $\sqrt[3]{5}$; (b) $1 + \sqrt[4]{3} + \sqrt[5]{7}$; (c) $3 + \sqrt[6]{1 + \sqrt{4}}$. In each case choose an x_0 by estimating the number sought and calculate a few terms of the sequence (11.5). If the sequence apparently diverges, try to reformulate the fixed point problem to obtain a sequence which appears to converge.

2. Equation (11.30) gives rise to the sequence $\{u_{11}^{(m)}\}$, where

$$u_{11}^{(m+1)} = \tfrac{3}{4}u_{11}^{(m)} + \frac{6}{u_{11}^{(m)}} - 4$$

for $u_{11}^{(0)} \neq 0$. Calculate a few terms of this sequence for $u_{11}^{(0)} = -18, -17, 1.3, 1.4$. Which initial values seem to yield a convergent sequence? To which of the solutions (11.31)?

3. Formulate the system (11.28) for $n = 4$, recalling that $u_{0j}, u_{4j}, j = 0, 1, 2, 3, 4$, and $u_{i0}, u_{i4}, i = 0, 1, 2, 3, 4$, are obtained from (11.20). If an electronic digital computer is available, investigate the solution of the resulting system of nine equations in nine unknowns by the method of interation (11.5). *Warning.* Do not expect convergence. Reduce the boundary conditions (11.20) by a factor of 10^{-2}. Does (11.5) appear to converge for the resulting problem?

4. Prove (11.16). Find a projection Π of R^2 into R with $\|\Pi\| > 1$.

5. Show that the orthogonal projection $p(x, y)$ defined by (6.35) can be expressed as $\Pi x = p(x, y)$, Π being a projection operator, $\|\Pi\| = 1$.

12. CONVERGENCE OF THE METHOD OF SUCCESSIVE SUBSTITUTIONS

The questions of existence, uniqueness, and construction or approximation of fixed points of operators which satisfy certain simple conditions are considered in this section, as well as the problem of error estimation for approximate solutions. The basis for the theory of fixed point equations developed here is the method of successive substitutions (11.5) which was introduced in the previous section.

Theorem 12.1. If F is a continuous operator in a Banach space X, and the sequence $\{x_m\}$ defined by

$$x_{m+1} = F(x_m), \qquad m = 0, 1, 2, \ldots, \tag{12.1}$$

for some $x_0 \in X$ converges to $x = x^*$, then x^* is a fixed point of the operator F, that is

$$x^* = F(x^*). \tag{12.2}$$

PROOF. This follows directly from (12.1) and Definition 8.1 of continuity of an operator, as

$$x^* = \lim_{m \to \infty} x_{m+1} = \lim_{m \to \infty} F(x_m) = F(x^*). \tag{12.3}$$

The essential ingredients of Theorem 12.1 are the continuity of F (at least in a ball which contains $\{x_m\}$ and x^*) and the convergence of the sequence $\{x_m\}$. Although the *existence* of a fixed point x^* may be established on the basis of Theorem 12.1, as well as the possibility of constructing it by the iterative process (12.1), nothing has been proved concerning the uniqueness of x^* or the error $\|x_m - x^*\|$ resulting from taking the indicated term of the convergent sequence as an approximation to x^*. These latter questions may be resolved by the introduction of the following idea.

Definition 12.1. An operator F in a Banach space X is called a *contraction mapping* of the closed ball $\bar{U}(x_0, r)$ if there exists a real number $\theta, 0 \leq \theta < 1$, such that

$$\|F(x) - F(y)\| \leq \theta \|x - y\| \tag{12.4}$$

for all $x, y \in \bar{U}(x_0, r)$.

It is obvious from this definition that a contraction mapping F maps the ball $\bar{U}(x_0, r)$ into the smaller ball $\bar{U}(F(x_0), \theta r)$. Furthermore, F is continuous at each point of $U(x_0, r)$, the interior of $\bar{U}(x_0, r)$, and, for any sequence $\{x_m\}$ contained in $\bar{U}(x_0, r)$ which converges to a point x^* of $\bar{U}(x_0, r)$,

$$\lim_{m \to \infty} F(x_m) = F(x^*). \tag{12.5}$$

A simple example of a contraction mapping is the scalar function

$$F(x) = x^2 + 2, \tag{12.6}$$

which maps the ball

$$\bar{U}(0, \tfrac{1}{2}) = \{x: |x| \le \tfrac{1}{2}\} \tag{12.7}$$

into the ball

$$\bar{U}(2, \tfrac{1}{4}) = \{y: |y - 2| \le \tfrac{1}{4}\}. \tag{12.8}$$

For the contraction mapping (12.6) of the ball (12.7), we have

$$\theta = \tfrac{1}{2} \tag{12.9}$$

in (12.4). The constant θ is called the *contraction factor* for F on $\bar{U}(x_0, r)$.

Theorem 12.2 (*The Contraction Mapping Principle*). If F is an operator in a Banach space X which is a contraction mapping of $\bar{U}(x_0, r)$ for

$$r \ge \frac{1}{1 - \theta} \|x_0 - F(x_0)\| = r_0, \tag{12.10}$$

where θ is the contraction factor for F on $\bar{U}(x_0, r)$, then:

1. F has a fixed point x^* in $\bar{U}(x_0, r_0)$.
2. x^* is the unique fixed point of F in $\bar{U}(x_0, r)$.
3. The sequence $\{x_m\}$ of successive approximations defined by (12.1) converges to x^*, with

$$\|x_m - x^*\| \le \theta^m r_0. \tag{12.11}$$

PROOF. First it will be shown that the sequence $\{x_m\}$ is contained in the ball $\bar{U}(x_0, r_0)$. As

$$x_1 = F(x_0), \tag{12.12}$$

it follows from (12.10) that

$$\|x_1 - x_0\| = (1 - \theta)r_0 < r_0, \tag{12.13}$$

so $x_1 \in \bar{U}(x_0, r_0)$. Assume that $x_0, x_1, \ldots, x_{n-1}, x_n \in \bar{U}(x_0, r_0)$, and that

$$\|x_n - x_0\| \le (1 - \theta^n)r_0 < r_0, \tag{12.14}$$

for some positive integer n. Then, since

$$\|x_{n+1} - x_n\| = \|F(x_n) - F(x_{n-1})\| \le \theta \|x_n - x_{n-1}\|, \tag{12.15}$$

successive applications of which give

$$\|x_{n+1} - x_n\| \le \theta^n \|x_1 - x_0\| = \theta^n(1 - \theta)r_0, \tag{12.16}$$

and

$$\|x_{n+1} - x_0\| \le \|x_n - x_0\| + \|x_{n+1} - x_n\|$$
$$\le (1 - \theta^n)r_0 + \theta^n(1 - \theta)r_0 = (1 - \theta^{n+1})r_0 < r_0, \tag{12.17}$$

the validity of (12.14) for all positive integers n has been established by mathematical induction. This proves the assertion that $\{x_m\}$ is contained in $\bar{U}(x_0, r_0)$. Now it will be shown that $\{x_m\}$ is a Cauchy sequence and, thus, has a limit point x^*, which belongs to $\bar{U}(x_0, r_0)$ by reason of the foregoing.

By the same reasoning as was used to obtain (12.14),

$$\|x_{m+p} - x_m\| \leq \frac{1 - \theta^p}{1 - \theta} \|x_{m+1} - x_m\|, \tag{12.18}$$

which, by (12.16), amounts to

$$\|x_{m+p} - x_m\| \leq (1 - \theta^p)\theta^m r_0. \tag{12.19}$$

It is obvious from this that the sequence $\{x_m\}$ is a Cauchy sequence by Definition 5.1. Hence its limit x^* exists in $\bar{U}(x_0, r_0)$ and is a fixed point of F by (12.5) and (12.1). To show the uniqueness of x^* in $\bar{U}(x_0, r)$, suppose that x^{**} is some other fixed point of F in $\bar{U}(x_0, r)$. Then, since F is a contraction mapping of $\bar{U}(x_0, r)$,

$$\|x^* - x^{**}\| = \|F(x^*) - F(x^{**})\| \leq \theta \|x^* - x^{**}\| < \|x^* - x^{**}\|, \tag{12.20}$$

which is impossible. The error bound (12.11) is obtained immediately by taking the limit as $p \to \infty$ in inequality (12.19). This completes the proof.

There is no essential restriction of Definition 12.1 or Theorem 12.2 to the case that r is finite; we may define $\bar{U}(x_0, \infty) = X$, the entire space, and introduce the idea of a *contraction mapping F of X*. For such an operator Theorem 12.2 guarantees the existence of a fixed point x^* of F in $\bar{U}(x_0, r_0)$ and its uniqueness in the entire space X.

To illustrate an application of Theorem 12.2, suppose that bounded linear operators K, K^{-1}, L and a point $y \in X$ are given. A solution of the linear equation

$$Lx = y \tag{12.21}$$

will be a fixed point of the (nonlinear) operator F defined by

$$F(x) = (I - KL)x + Ky, \tag{12.22}$$

and conversely. If

$$\theta = \|I - KL\| < 1, \tag{12.23}$$

Theorem 12.2 states that an x^* exists such that

$$x^* = (I - KL)x^* + Ky; \tag{12.24}$$

that is, $Lx^* = y$ exists in the ball $\bar{U}(x_0, r_0)$, where

$$r_0 = (1 - \|I - KL\|)^{-1}\|K(Lx_0 - y)\| \tag{12.25}$$

for any $x_0 \in X$ and x^* is unique in X. The error bound

$$\|x_m - x^*\| \leq \frac{\|I - KL\|^m}{1 - \|I - KL\|} \|K(Lx_0 - y)\| \tag{12.26}$$

is obtained for

$$x_m = (I - KL)x_{m-1} + Ky, \qquad m = 1, 2, 3, \ldots. \tag{12.27}$$

These results are essentially the same as were obtained on the basis of Theorem 9.1, condition (12.23) being exactly (9.13). Of course, Theorem 12.2 only establishes the sufficiency of the existence of K, K^{-1} and (12.23) for the unique solvability of (12.21), whereas Theorem 9.1 shows that these conditions are necessary as well as sufficient.

As another example, the previously cited real operator F defined by

$$F(x) = \frac{1}{2}\left(x + \frac{3}{x}\right) \tag{12.28}$$

is considered. It is simple to derive that

$$F(x) - F(y) = \frac{1}{2}\left(1 - \frac{3}{xy}\right)(x - y) \tag{12.29}$$

for $xy \neq 0$. In the ball $\bar{U}(2, \frac{1}{2})$

$$|F(x) - F(y)| \leq 0.26\,|x - y|, \tag{12.30}$$

so F is a contraction mapping of $\bar{U}(2, \frac{1}{2})$ with $\theta = 0.26$. For $x_0 = 2, x_1 = 1.75$ and

$$r_0 = \frac{1}{1 - 0.26}\,0.25 < 0.338 < 0.5 = r, \tag{12.31}$$

so that Theorem 12.2 guarantees the existence of a fixed point x^* of F in $\bar{U}(2, 0.388)$ to which the sequence $\{x_m\}$ defined by

$$x_0 = 2, \qquad x_{m+1} = \frac{1}{2}\left(x_m + \frac{3}{x_m}\right), \qquad m = 0, 1, 2, \ldots, \tag{12.32}$$

converges, with

$$|x^* - x_m| \leq (0.26)^m 0.338. \tag{12.33}$$

Actually it is shown later that the sequence (12.32) converges much faster to $x^* = \sqrt{3}$ than indicated by (12.33). However, the above results give a rigorous demonstration of the convergence of (12.32), in this case for $x_0 = 2$. It was asserted earlier that (12.32) converges for $x_0 = 1$. This is easy to prove, for if $x_0 = 1$ then

$$x_1 = \frac{1}{2}(1 + \frac{3}{1}) = 2, \tag{12.34}$$

which obviously gives rise to a sequence that can be identified with (12.32) except for its first term.

Recalling that a positive integral power F^n of an operator F is defined by

$$F = F^{n-1}, \qquad F^0 = I, \qquad n = 1, 2, \ldots, \qquad (12.35)$$

it is evident that a fixed point x^* of F is also a fixed point of the operator F^n for any positive integer n. The sequence $\{x_m\}$ defined by (12.1) can also be defined by

$$x_m = F^m(x_0), \qquad m = 0, 1, 2, \ldots. \qquad (12.36)$$

Definition 12.2. A fixed point x^* of an operator F is said to be *accessible from x_0* if

$$x^* = \lim_{m \to \infty} F^m(x_0). \qquad (12.37)$$

If x^* is accessible from x_0, it is accessible from any point x_m of the sequence $\{x_m\}$ defined by (12.36) and also from any point x_{-k} such that

$$x_0 = F^k(x_{-k}) \qquad (12.38)$$

for some positive integer k.

The set of points x_0 for which (12.37) is satisfied is called the *region of accessibility* of the fixed point x^* of F. For each fixed point x^* of F, there is a nonempty region of accessibility, since $x_0 = x^*$ satisfies (12.37). For the scalar operator $F(x) = x^2$ the region of accessibility of the fixed point $x^* = 0$ is the set of x_0 with $|x_0| < 1$, whereas the fixed point $x^* = 1$ has a region of accessibility that consists only of the two points $x_0 = 1$, $x_0 = -1$.

Theorem 12.3. If the hypotheses of Theorem 12.2 are satisfied, the unique fixed point x^* of F in $\bar{U}(x_0, r)$ is accessible from any point of the closed ball $\bar{U}(x_0, r_0)$.

PROOF.   For $x \in \bar{U}(x_0, r_0)$

$$\|F(x) - F(x_0)\| = \|F(x) - x_1\| \leq \theta \|x - x_0\|, \qquad (12.39)$$

and, since

$$\|x - x_0\| \leq r_0 = \frac{\|x_1 - x_0\|}{1 - \theta}, \qquad (12.40)$$

it follows that

$$\|F(x) - x_1\| \leq \frac{\theta}{1 - \theta} \|x_1 - x_0\| = \theta r_0. \qquad (12.41)$$

Thus F maps the ball $\bar{U}(x_0, r_0)$, which contains x^*, into the smaller ball $\bar{U}(x_1, \theta r_0)$, which also contains x^* by (12.11). We also have that $\bar{U}(x_1, \theta r_0) \subset \bar{U}(x_0, r_0)$; that is, $\bar{U}(x_1, \theta r_0)$ is a subset of $\bar{U}(x_0, r_0)$. This is true, for if $x \in \bar{U}(x_1, \theta r_0)$ then

$$\|x - x_0\| \leq \|x - x_1\| + \|x_1 - x_0\| \leq \theta r_0 + (1 - \theta)r_0 = r_0. \quad (12.42)$$

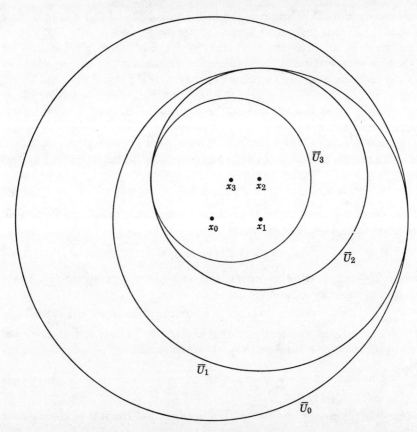

Figure 12.1 $\theta = 0.75$, $\bar{U}_m = \bar{U}(x_m, \theta^m r_0)$, $m = 0, 1, 2, 3$.

See Figure 12.1, which illustrates the case $\theta = 0.75$. Continuing in this way we find that F maps $\bar{U}(x_m, \theta^m r_0)$ into $\bar{U}(x_{m+1}, \theta^{m+1} r_0)$, and for

$$x \in \bar{U}(x_{m+1}, \theta^{m+1} r_0)$$

$$\|x^* - x\| \leq \|x^* - x_{m+1}\| + \|x - x_{m+1}\| \leq 2\theta^{m+1} r_0 \qquad (12.43)$$

by (12.11). Since, for any $x_0' \in U(x_0, r_0)$,

$$x_{m+1}' = F(x_m'), \qquad m = 0, 1, 2, \ldots, \qquad (12.44)$$

will be an element of $\bar{U}(x_{m+1}, \theta^{m+1} r_0)$, the convergence of the sequence $\{x_m'\}$ to x^* follows from (12.43). This proves the theorem.

It is possible to interpret the result of Theorem 12.3 in a somewhat different way. Any sequence $\{\tilde{x}_m\}$ such that

$$\tilde{x}_m \in \bar{U}(x_m, \theta^m r_0), \qquad m = 0, 1, 2, \ldots, \qquad (12.45)$$

will converge to x^*. This gives rise to the possibility of using approximate values for $F(x)$ in generating successive approximations to x^*. This can be important in applications, since the precise evaluation of $F(x)$ may be laborious or unattainable. In addition, the values of $x_1, x_2, \ldots, x_n$ obtained in the early stages of the computation may not be particularly close to x^*, so it might be more efficient to invest computational effort in more iterations rather than in excessively accurate evaluation of the initial terms of the sequence.

To arrive at a formal statement of these ideas suppose that F is a contraction mapping of the ball $\bar{U}(x_0, r)$ and consider the sequence $\{\tilde{x}_m\}$ defined by

$$\tilde{x}_0 = x_0, \qquad \tilde{x}_{m+1} = F(\tilde{x}_m) + \delta_m, \qquad m = 0, 1, 2, \ldots. \tag{12.46}$$

The sequence (12.46) can be regarded as arising from evaluating $F(\tilde{x}_m)$ with an error equal to δ_m. Suppose that

$$\|\delta_m\| \leq \lambda^m \delta, \qquad m = 0, 1, 2, \ldots, \tag{12.47}$$

where $0 \leq \lambda < 1$. In other words, the intention is to evaluate $F(\tilde{x}_m)$ with increasing accuracy as m increases.

Theorem 12.4. If F is a contraction mapping of the ball $U(x_0, r)$, and (12.47) is satisfied, the sequence $\{\tilde{x}_m\}$ defined by (12.46) converges to the unique fixed point x^* of F in $\bar{U}(x_0, r)$, provided that

$$r \geq r_0 + \frac{\delta}{1 - \theta} = \tilde{r}_0, \tag{12.48}$$

where θ is the contraction factor for F on $\bar{U}(x_0, r)$. The rate of convergence is given by

$$\|\tilde{x}_m - x^*\| \leq \theta^m r_0 + m\delta \, (\max \{\lambda, \theta\})^{m-1}. \tag{12.49}$$

PROOF. Suppose that $\tilde{x}_0 = x_0, \tilde{x}^1, \ldots, \tilde{x}_{m-1}, \tilde{x}_m$ belong to the ball $\bar{U}(x_0, \tilde{r}_0)$. By comparison of (12.46) with (12.1)

$$\|\tilde{x}_{m+1} - x_{m+1}\| = \|F(\tilde{x}_m) - F(x_m) + \delta_m\|$$
$$\leq \theta \|\tilde{x}_m - x_m\| + \|\delta_m\|$$
$$\leq \theta^2 \|\tilde{x}_{m-1} - x_{m-1}\| + \theta \|\delta_{m-1}\| + \|\delta_m\|$$
$$\vdots$$
$$\leq \sum_{i=0}^{m} \theta^{m-i} \|\delta_i\|$$
$$< \frac{\delta}{1 - \theta}. \tag{12.50}$$

As
$$\|\tilde{x}_{m+1} - x_0\| \leq \|x_{m+1} - x_0\| + \|\tilde{x}_{m+1} - x_{m+1}\|$$
$$\leq r_0 + \frac{\delta}{1 - \theta} = \tilde{r}_0, \tag{12.51}$$

we have that $\tilde{x}_{m+1} \in \bar{U}(x_0, \tilde{r}_0)$, and consequently the sequence $\{\tilde{x}_m\}$ belongs to $\bar{U}(x_0, \tilde{r}_0)$. Returning to inequality (12.50), we obtain the more precise estimate

$$\|\tilde{x}_m - x_m\| \leq \delta \sum_{i=0}^{m-1} \theta^{m-i-1}\lambda^i, \tag{12.52}$$

by use of (12.47). For
$$\eta = \max \{\theta, \lambda\} \tag{12.53}$$

(12.52) becomes
$$\|\tilde{x}_m - x_m\| \leq m\delta\eta^{m-1}, \tag{12.54}$$

which tends to zero as $m \to \infty$, since $0 \leq \eta < 1$. As
$$\|\tilde{x}_m - x^*\| \leq \|x_m - x^*\| + \|\tilde{x}_m - x_m\|$$
$$\leq \theta^m r_0 + m\delta\eta^{m-1}, \tag{12.55}$$

the convergence of the sequence $\{\tilde{x}_m\}$ to x^* and the error bound (12.49) are obtained. The existence of x^* as the limit of the sequence $\{x_m\}$ defined by (12.1) and its uniqueness in $\bar{U}(x_0, r)$ follow, of course, from Theorem 12.2. This proves the theorem.

The rule of thumb for calculating with scalar equations that F should be evaluated with one more decimal place of accuracy at each iteration corresponds to the case $\lambda = 0.1, \delta = 1.0$. Another possible application of Theorem 12.4 occurs in connection with the conversion of analytic fixed point problems into arithmetic fixed point problems. For example, if F involves integral or differential operators, rather crude rules of numerical integration or differentiation [9] can be used in the early stages of the calculation, with more refined methods being introduced when greater accuracy is required. Thus in the solution of the partial differential equation $\Delta u = u^2$ [see (11.18)], with the boundary conditions (11.20), we could use the solution of the difference equations (11.28) for $h = \frac{1}{2}$ to provide initial values for $h = \frac{1}{4}$, and so on.

It is possible to establish the convergence of the method of iteration under less restrictive conditions than the contraction mapping principle [5]. To this end the following idea is introduced.

Definition 12.3. If P is an operator in a Banach space X, the quantity

$$M(P, x_0, r) = \sup_{x \neq y} \frac{\|P(x) - P(y)\|}{\|x - y\|}, \qquad x, y \in \bar{U}(x_0, r), \tag{12.56}$$

is called the *bound* of P on the ball $\bar{U}(x_0, r)$.

Once the ball $\bar{U}(x_0, r)$ has been specified, the abbreviated notation

$$\|P\| = M(P, x_0, r) \tag{12.57}$$

will be used. If P is linear, this is consistent with the definition already given. [see (8.20)]. If $\|P\|$ is finite, P is obviously continuous in the open ball $U(x_0, r)$, and (12.5) holds for any $x^* \in \bar{U}(x_0, r)$. The operator P will be a contraction mapping of $\bar{U}(x_0, r)$ if $\|P\| < 1$.

Theorem 12.5.   Suppose that in a ball $U(x_0, r)$, the series

$$R = \sum_{n=0}^{\infty} \|F^n\| \tag{12.58}$$

converges, and

$$r \geq R \|F(x_0) - x_0\| = r_0. \tag{12.59}$$

Then F has a fixed point x^* in the ball $\bar{U}(x_0, r_0)$, which is unique in the ball $\bar{U}(x_0, r)$, and the sequence $\{x_m\}$ defined by (12.1) converges to x^*, with

$$\|x_m - x^*\| \leq \left(R - \sum_{k=0}^{m-1} \|F^k\|\right) \|F(x_0) - x_0\|. \tag{12.60}$$

PROOF.   The sequence $\{x_m\}$ is contained in the ball $U(x_0, r_0)$, since

$$\|x_m - x_0\| = \sum_{i=1}^{m} \|x_i - x_{i-1}\|$$

$$\leq \sum_{i=1}^{m} \|F^{i-1}(x_1) - F^{i-1}(x_0)\|$$

$$\leq \sum_{i=1}^{m} \|F^{i-1}\| \cdot \|x_1 - x_0\|$$

$$\leq R \|F(x_0) - x_0\| = r_0. \tag{12.61}$$

By the same reasoning

$$\|x_{m+p} - x_m\| \leq \sum_{i=m}^{m+p-1} \|F^i\| \cdot \|F(x_0) - x_0\|, \tag{12.62}$$

which shows that $\{x_m\}$ is a Cauchy sequence as a consequence of the convergence of the series (12.58). Hence $\{x_m\}$ has a limit $x^* \in \bar{U}(x_0, r_0)$, which is a fixed point of F by (12.5). The uniqueness of x^* in $\bar{U}(x_0, r)$ follows from the fact that F^p is a contraction mapping of $\bar{U}(x_0, r)$ for some positive integer p, once again from the convergence of the series (12.58). Since any fixed point x^{**} of F in $\bar{U}(x_0, r)$ will be a fixed point of F^p,

$$\|x^* - x^{**}\| = \|F^p(x^*) - F^p(x^{**})\| \leq \|F^p\| \cdot \|x^* - x^{**}\|, \tag{12.63}$$

which, since $\|F^p\| < 1$, will be impossible if $x^* \neq x^{**}$. The error bound (12.60) follows at once by taking the limit as $p \to \infty$ in (12.62) and recalling (12.58). This completes the proof.

Theorem 12.5 is essentially the same as the fixed point theorem of Weissinger [5], who gave applications to a number of examples. One simple application is to obtain the standard theorem on the existence and uniqueness of a solution $y \in C'[0, T]$ of the nonlinear ordinary differential equation

$$\frac{dy}{dt} = f(t, y(t)), \qquad 0 \leq t \leq T,$$
$$y(0) = y_0. \tag{12.64}$$

Equation 12.64 may be put into the form of the fixed point problem [recall (11.2)]

$$y(t) = y_0 + \int_0^t f(s, y(s))\, ds, \qquad 0 \leq t \leq T, \tag{12.65}$$

where, if $f(s, y)$ is continuous and satisfies the *Lipschitz condition*,

$$\max_{[0,T]} |f(s, x_1(s)) - f(s, x_2(s))| \leq M \, \|x_1 - x_2\| \tag{12.66}$$

for $x_1, x_2 \in C[0, T]$, then the right-hand side of (12.65) will define an operator F from $C[0, T]$ into itself. By the same analysis used for the linear Volterra integral equation (10.15) it is not difficult to show that

$$\|F^n\| \leq \frac{M^n T^n}{n!}, \tag{12.67}$$

which guarantees the convergence of the series (12.58) and thus the convergence of the *Picard sequence*

$$y_0(t) = y_0, \qquad y_{m+1}(t) = y_0 + \int_0^t f(s, y_m(s))\, ds,$$
$$0 \leq t \leq T, \quad m = 0, 1, 2, \ldots \tag{12.68}$$

to a solution $y(t)$ of (12.65) [hence of (12.64)], which exists in a ball of radius

$$r_0 = e^{MT} \, \|y_1 - y_0\|, \tag{12.69}$$

and is unique in the entire space. The error bound

$$\|y(t) - y_m(t)\| \leq \left(e^{MT} - \sum_{k=0}^{m-1} \frac{M^k T^k}{k!} \right) \|y_1 - y_0\| \tag{12.70}$$

follows from (12.60). This result may be extended to the case that (12.66) holds in a ball $\bar{U}(y_0, r)$, with $r \geq r_0$ defined by (12.69). If T is chosen sufficiently small, this condition is possible to satisfy once $r > 0$ is known (see Problem 4), provided, of course, that $r > \|y_1 - y_0\|$.

Theorem 12.5 has somewhat the same relation to Theorem 12.2 that Theorem 10.2 has to Theorem 9.1. However, the theorems on linear operators were *global*, stating existence and uniqueness of x^* in the entire space, whereas the theorems on nonlinear operators are *local*, generally asserting only that fixed points exist and are unique in some ball.

EXERCISES 12

1. For the scalar equation $x = F(x)$, a fixed point x^* in the real case will be the abscissa of an intersection point of the straight line $y = x$, with the curve $y = F(x)$. Give a geometric procedure for constructing the points $x_{m+1} = F(x_m)$, $m = 0, 1, 2, \ldots$. Sketch examples of convergent and divergent sequences.
2. Give a geometric interpretation of the contraction mapping principle in the real scalar case. Analyze (11.14) graphically in the neighborhood of the fixed point $x^* = \sqrt{3}$. Analyze (11.30) graphically in the neighborhood of each of its fixed points. $x^* = 1.380832\ldots$, $-17.380832\ldots$.
3. Suppose that the operator F in R_∞^n defined by

$$f_i = f_i(\xi_1, \xi_2, \ldots, \xi_n), \qquad i = 1, 2, \ldots, n,$$

is a contraction mapping of the ball $\overline{U}(x_0, 10)$, $\|x_1 - x_0\| = 1$, and $\theta = 0.75$. Also, assume that $F(x)$ can be calculated to an accuracy of 2^{-p} in $100p$ msec by a given computer program. Determine values of δ and λ on the basis of Theorem 12.4 which assure achieving the accuracy

$$\|\tilde{x}_m - x^*\| \leq 2^{-24}$$

with the least expenditure of computer time.
4. Apply Theorem 12.5 to the ordinary differential equation

$$\frac{dy}{dt} = 1 + y^2, \qquad y(0) = 0, \qquad 0 \leq t \leq T.$$

Determine a value of T for which existence and uniqueness of a solution is guaranteed and obtain the terms $y_i(t)$, $i = 1, 2, 3, 4$, of the Picard sequence (12.68).
5. Give sufficient conditions for the solvability of the quadratic scalar equation

$$x = a + bx^2$$

by the method of successive substitutions, starting from $x_0 = 0$. Derive explicit formulas for x_1, x_2, x_3.

13. A NONLINEAR INTEGRAL EQUATION

The ideas in the previous sections will be applied to the simple nonlinear integral equation

$$H(\mu) = 1 + \tfrac{1}{2}\varpi_0\mu\, H(\mu) \int_0^1 \frac{H(\mu')}{\mu + \mu'}\, d\mu', \qquad 0 \leq \mu \leq 1, \qquad (13.1)$$

which arises in the theory of radiative transfer [10]. In (13.1), $H(\mu)$ is the unknown function, which will be sought in $C[0, 1]$. The quantity ϖ_0 ("pi-nought"; the symbol ϖ is a script pi, not a fancy omega) is a parameter called the *albedo* that will have a value in the range

$$0 \leq \varpi_0 \leq 1. \tag{13.2}$$

The physical background of (13.1) is fairly elaborate. This equation was developed by Ambartsumian [11] and Chandrasekhar [10] to solve the problem of determination of the angular distribution of the radiant flux emerging from a semi-infinite, plane-parallel, isotropic atmosphere. Explicit definitions of these terms may be found in the literature [10]; for the present purposes (13.1) is regarded as a typical example of a class of nonlinear computational problems. In particular, (13.1) is the prototype of the equation

$$H(\mu) = 1 + \mu H(\mu) \int_0^1 \frac{\Psi(\mu')}{\mu + \mu'} H(\mu') \, d\mu', \tag{13.3}$$

for more general laws of scattering [10], where $\Psi(\mu)$ is an even polynomial in μ, with

$$\int_0^1 \Psi(\mu) \, d\mu \leq \tfrac{1}{2}. \tag{13.4}$$

Integral equations of the form (13.3) also arise in other studies [13]. Consequently a direct attack on (13.1) may yield insight into a number of related problems.

It has been shown [12] that (13.1) has the explicit solution

$$H(\mu) = \exp\left[-\frac{\mu}{\pi} \int_0^{\pi/2} \frac{\ln(1 - \varpi_0 \theta \cot \theta)}{\sin^2 \theta + \mu^2 \cos^2 \theta} \, d\theta \right], \qquad 0 \leq \mu \leq 1. \tag{13.5}$$

The evaluation of (13.5) requires the use of numerical integration, which can be done to any desired accuracy [14]. For the particular equation being considered, the known solution (13.5) will allow the comparison of theoretical error estimates for various methods of approximate solution with the actual error.

The form of (13.1) suggests the direct iteration

$$H_{m+1}(\mu) = 1 + \tfrac{1}{2}\varpi_0 \mu H_m(\mu) \int_0^{} \frac{H_m(\mu')}{\mu + \mu'} \, d\mu', \qquad 0 \leq \mu \leq 1, \quad m = 0, 1, 2, \ldots, \tag{13.6}$$

with

$$H_0(\mu) = 1, \qquad 0 \leq \mu \leq 1. \tag{13.7}$$

In the following this function will sometimes be denoted simply by 1. We obtain immediately that

$$H_1(\mu) = 1 + \tfrac{1}{2}\varpi_0 \mu \ln\left(1 + \frac{1}{\mu}\right), \qquad 0 \leq \mu \leq 1. \tag{13.8}$$

The exact calculation of $H_2(\mu)$, however, involves evaluation of the integral

$$\int_0^1 \frac{\mu'}{\mu + \mu'} \ln\left(1 + \frac{1}{\mu'}\right) d\mu', \qquad 0 \le \mu \le 1, \tag{13.9}$$

which leads to analytic complexities. However, (13.7) and (13.8) provide some of the information necessary for the application of Theorem 12.2. We have

$$\|H_1 - H_0\| = \max_{[0,1]} |H_1(\mu) - H_0(\mu)| = \tfrac{1}{2}\varpi_0 \ln 2, \tag{13.10}$$

or

$$\|H_1 - H_0\| < 0.34658\varpi_0. \tag{13.11}$$

If the operator F in $C[0, 1]$ defined by

$$F(H)(\mu) = 1 + \tfrac{1}{2}\varpi_0\mu H(\mu) \int_0^1 \frac{H(\mu')}{\mu + \mu'} d\mu', \qquad 0 \le \mu \le 1, \tag{13.12}$$

is a contraction mapping of the ball $\bar{U}(1, r)$ with the contraction factor θ, and

$$r \ge \left(\frac{1}{1 - \theta}\right)\tfrac{1}{2}\varpi_0 \ln 2, \tag{13.13}$$

then the existence of a solution $H(\mu)$ of (13.1) and the convergence of the sequence $\{H_m\}$ generated by (13.6) to it follow from the contraction mapping principle.

To determine the value of θ note that

$$F(x)(\mu) - F(y)(\mu) = \tfrac{1}{2}\varpi_0\mu\left[x(\mu) \int_0^1 \frac{x(\mu')}{\mu + \mu'} d\mu' - y(\mu) \int_0^1 \frac{y(\mu')}{\mu + \mu'} d\mu' \right]$$

$$= \tfrac{1}{4}\varpi_0\mu\left\{ [x(\mu) + y(\mu)] \int_0^1 \frac{1}{\mu + \mu'} [x(\mu') - y(\mu')] \, d\mu \right.$$

$$\left. + [x(\mu) - y(\mu)] \int_0^1 \frac{1}{\mu + \mu'} [x(\mu') + y(\mu')] \, d\mu' \right\},$$

$$0 \le \mu \le 1. \tag{13.14}$$

If $x, y \in \bar{U}(1, r)$, then

$$\|x + y\| \le 2(1 + r), \tag{13.15}$$

and

$$\max_{[0,1]} \left| \int_0^1 \frac{\mu}{\mu + \mu'} |x(\mu') + y(\mu')| \, d\mu' \right| \le 2(1 + r) \ln 2. \tag{13.16}$$

It follows from (13.14) that

$$\|F(x) - F(y)\| \le \varpi_0(1 + r)(\ln 2) \|x - y\|, \tag{13.17}$$

and F is a contraction mapping of $\bar{U}(1, r)$, provided that

$$\theta = \varpi_0(1 + r) \ln 2 < 1. \tag{13.18}$$

From (13.13) and (13.18) the hypotheses of Theorem 12.2 will be satisfied for a given value of ϖ_0 if the inequality

$$r \geq \frac{\frac{1}{2}\varpi_0 \ln 2}{1 - \varpi_0(1 + r) \ln 2} \tag{13.19}$$

holds. For convenience set

$$a = \varpi_0 \ln 2. \tag{13.20}$$

Inequality (13.19) becomes

$$ar^2 - (1 - a)r + \tfrac{1}{2}a \leq 0, \tag{13.21}$$

which has the formal solution

$$\frac{1 - a - \sqrt{(1 - a)^2 - 2a^2}}{2a} \leq r \leq \frac{1 - a + \sqrt{(1 - a)^2 - 2a^2}}{2a}, \tag{13.22}$$

if we assume that the radicands are nonnegative. The maximum value of a for which this will be true is determined from

$$(1 - a)^2 - 2a^2 = 0 \tag{13.23}$$

to be

$$a = \sqrt{2} - 1. \tag{13.24}$$

For this value of a, (13.22) gives

$$r = \frac{\sqrt{2}}{2}. \tag{13.25}$$

From (13.24) and (13.25),

$$\theta = (\sqrt{2} - 1)\left(1 + \frac{\sqrt{2}}{2}\right) = \frac{\sqrt{2}}{2} < 1, \tag{13.26}$$

so that (13.18) is satisfied.

Thus $H_0(\mu) = 1$, $0 \leq \mu \leq 1$, is a satisfactory initial approximation to a solution of (13.1) if

$$0 \leq \varpi_0 \leq \frac{\sqrt{2} - 1}{\ln 2} = 0.59758\cdots. \tag{13.27}$$

For each ϖ_0 in this range Theorem 12.2 guarantees the existence of a solution in the ball $\bar{U}(1, r_0{}^U)$, and its uniqueness in $\bar{U}(1, r_0{}^E)$, where

$$r_0{}^E = \frac{1 - a - \sqrt{(1 - a)^2 - 2a^2}}{2a} \tag{13.28}$$

and

$$r_0^U = \frac{1 - a + \sqrt{(1 - a)^2 - 2a^2}}{2a}, \tag{13.29}$$

with $a = \varpi_0 \ln 2$, from (13.20) and (13.22). These estimates yield the additional information that

$$\|H\| = \max_{[0,1]} |H(\mu)| \leq 1 + r_0^E, \tag{13.30}$$

and the error bound

$$\|H - H_1\| = \max_{[0,1]} |H(\mu) - H_1(\mu)| \leq \varpi_0 (\ln 2)(1 + r_0^E) r_0^E \tag{13.31}$$

for the solution $H(\mu)$ of (13.1) and the approximation $H_1(\mu)$ to it are given by (13.8). Inequality (13.31) follows directly from (13.18) and (12.11).

From the standpoint of the original problem, the limitation on ϖ_0 imposed by (13.27) is unsatisfactory, since solutions of (13.1) are desired over the entire range $0 \leq \varpi_0 \leq 1$. This means that other initial approximations $H_0(\mu)$ than (13.7) must be sought for $(\sqrt{2} - 1)/\ln 2 < \varpi_0 \leq 1$. One idea that suggests itself is to use the solution $H(\mu)$ of (13.1) for, say $\varpi_0 = 0.5$, as an initial approximation for larger values of ϖ_0. This process could conceivably be repeated until a satisfactory initial approximation is found for $\varpi_0 = 1$. Such tactics are known as *continuation*.

Because of the analytic difficulties in dealing directly with (13.1), it is convenient to construct a corresponding arithmetic model. This is done by introducing a *numerical integration rule* of the form

$$\int_0^1 f(s) \, ds \simeq \sum_{i=1}^n w_i f(s_i). \tag{13.32}$$

In (13.32) the points s_i, $0 \leq s_i \leq 1$, $i = 1, 2, \ldots, n$, are called the *nodes* and the numbers w_i, $i = 1, 2, \ldots, n$, are called the *weights* of the given rule. The number n is sometimes called the *order* of the rule.

On the basis of (13.32) the integral transform in (13.1) can be approximated as

$$\int_0^1 \frac{H(\mu')}{\mu + \mu'} \, d\mu' \simeq \sum_{j=1}^n \frac{w_j}{\mu + \mu_j} H(\mu_j), \qquad 0 \leq \mu \leq 1. \tag{13.33}$$

Thus for

$$b_{ij} = \frac{\mu_i w_j}{\mu_i + \mu_j}, \qquad i, j = 1, 2, \ldots, n, \tag{13.34}$$

(13.1) may be approximated by the arithmetic fixed point problem

$$\xi_i = 1 + \tfrac{1}{2}\varpi_0 \xi_i \sum_{j=1}^n b_{ij} \xi_j, \qquad i = 1, 2, \ldots, n, \tag{13.35}$$

where

$$\xi_i \cong H(\mu_i), \qquad i = 1, 2, \ldots, n. \tag{13.36}$$

Equations 13.35 may be written as

$$x = 1 + \tfrac{1}{2}\varpi_0 x \bigcirc Bx, \tag{13.37}$$

where $x = (\xi_1, \xi_2, \ldots, \xi_n)$, 1 denotes the vector $(1, 1, \ldots, 1)$, $B = (b_{ij})$ is the matrix defined by (13.34), and $\bigcirc$ stands for component-by-component multiplication of vectors, that is,

$$x \bigcirc y = (\xi_1\eta_1, \xi_2\eta_2, \ldots, \xi_n\eta_n) \tag{13.38}$$

for $x = (\xi_1, \xi_2, \ldots, \xi_n), y = (\eta_1, \eta_2, \ldots, \eta_n)$. The iteration formula (13.6) takes the form

$$x_{m+1} = 1 + \tfrac{1}{2}\varpi_0 x_m \bigcirc Bx_m \tag{13.39}$$

or

$$\xi_i^{(m+1)} = 1 + \tfrac{1}{2}\varpi_0 \xi_i^{(m)} \sum_{j=1}^{n} b_{ij}\xi_j^{(m)}, \qquad i = 1, 2, \ldots, n, \tag{13.40}$$

for some $x_0 = (\xi_1^{(0)}, \xi_2^{(0)}, \ldots, \xi_n^{(0)})$.

As an actual illustration, a *gaussian integration rule* [9] of order 9 will be used. The nodes and weights for this rule are given in Table 13.1. This rule has the property that it will integrate any polynomial on [0, 1] exactly if the degree of the polynomial does not exceed 17 and the exact values of the nodes and weights are used [9].

TABLE 13.1

Nodes and weights for the gaussian
integration rule of order nine

i	s_i	w_i
1	0.0159199	0.0406372
2	0.0819844	0.0903241
3	0.1933143	0.1303053
4	0.3378733	0.1561735
5	0.5000000	0.1651197
6	0.6621267	0.1561735
7	0.8066857	0.1303053
8	0.9180156	0.0903241
9	0.9840801	0.0406372

For a problem of this size, the use of an electronic digital computer is indicated [15]. The matrix $B = (b_{ij})$, $i, j = 1, 2, \ldots, 9$, defined by (13.34) is given in Table 13.2. In R_∞^9,

$$\|B\| = 0.69004017, \tag{13.41}$$

TABLE 13.2

The matrix $B = (b_{ij})$, $i,j = 1, 2, \ldots, 9$

0.02031857	0.01468729	0.00991447	0.00702745	0.00509514	0.00366680	0.00252179	0.00153966	0.00064693
0.03402929	0.04516203	0.03880512	0.03049554	0.02326048	0.01720682	0.01202133	0.00740516	0.00312515
0.03754525	0.06342542	0.06515264	0.05683598	0.04603972	0.03529240	0.02518987	0.01571175	0.00667214
0.03880861	0.07268677	0.08288348	0.07808674	0.06658469	0.05276685	0.03846606	0.02430000	0.01038631
0.03938324	0.07760010	0.09397274	0.09319636	0.08255985	0.06719297	0.04986099	0.03184876	0.01369103
0.03968307	0.08037240	0.10085864	0.10340664	0.09407765	0.07808674	0.05874039	0.03784848	0.01634483
0.03985074	0.08199123	0.10511542	0.11007115	0.10193706	0.08577196	0.06515264	0.04224682	0.01830582
0.03994449	0.08291893	0.10763887	0.11415795	0.10689759	0.09073214	0.06935826	0.04516205	0.01961288
0.03999025	0.08337783	0.10891070	0.11625767	0.10948938	0.09335840	0.07160671	0.04673064	0.02031859

TABLE 13.3
Solutions ξ_i of (13.40)

i	μ_i	$\varpi_0 = 0.1$	$\varpi_0 = 0.2$	$\varpi_0 = 0.3$	$\varpi_0 = 0.4$	$\varpi_0 = 0.5$	$\varpi_0 = 0.6$	$\varpi_0 = 0.7$	$\varpi_0 = 0.8$	$\varpi_0 = 0.9$	$\varpi_0 = 1.0$
1	0.0159199	1.00333256	1.00680007	1.01042392	1.01423279	1.01826693	1.02258650	1.02728988	1.03256294	1.03886882	1.05078677
2	0.0819844	1.01089700	1.02250493	1.03494913	1.04839844	1.06309081	1.07938298	1.09785983	1.11962814	1.14744584	1.20849350
3	0.1933143	1.01829896	1.03817916	1.05995689	1.08406434	1.11111939	1.14206177	1.17845465	1.22329853	1.28417453	1.43699152
4	0.3378733	1.02435469	1.05121662	1.08114160	1.11490166	1.15362021	1.19903838	1.25411203	1.32464171	1.42564767	1.71327104
5	0.5000000	1.02892234	1.06117664	1.09755914	1.13919212	1.18773523	1.24580731	1.31794528	1.41326285	1.55603409	2.01196827
6	0.6621267	1.03220522	1.06840231	1.10959652	1.15722114	1.21342326	1.28164112	1.36793413	1.48472709	1.66599375	2.30478155
7	0.8066857	1.03445865	1.07339473	1.11797621	1.16988219	1.23165185	1.30739610	1.40445453	1.53812001	1.75110912	2.56295413
8	0.9180156	1.03589121	1.07658239	1.12335356	1.17805497	1.24350204	1.32428636	1.42867807	1.57409997	1.80995644	2.76051640
9	0.9840801	1.03664375	1.07826123	1.12619412	1.18238741	1.24981064	1.33332583	1.44173219	1.59367977	1.84249973	2.87735197

which is slightly smaller than the bound $\ln 2 = 0.69315\ldots$ for the integral transform in (13.1). For

$$a = \varpi_0 \,\|B\| = 0.69004017\varpi_0, \qquad (13.42)$$

the previous relationships (13.24) and (13.25) may be used to obtain the range of ϖ_0 for which $x_0 = (1, 1, \ldots, 1)$ is a suitable initial approximation to the solution of (13.37) in the ball $\bar{U}(1, r_0{}^E)$, with $r_0{}^E$ given by (13.42) and (13.28). The uniqueness of this solution in $\bar{U}(1, r_0{}^U)$, where $r_0{}^U$ is given by (13.42) and (13.29), is also guaranteed by Theorem 12.2.

The solutions of (13.40) obtained by iteration for

$$\varpi_0 = 0.1(0.1)1.0 \qquad (13.43)$$

are shown in Table 13.3. The notation

$$y = y_0(h)y_N \qquad (13.44)$$

used in (13.43) is a convenient abbreviation for the set of values

$$y = y_0, y_0 + h, y_0 + 2h, \ldots, y_N - h, y_N. \qquad (13.45)$$

In performing these calculations, the method of continuation was used, the solution of (13.40) for $\varpi_0 = 0.1$ being employed as the initial approximation for $\varpi_0 = 0.2$, and so on. The number of iterations required to obtain convergence to eight decimal places for each value of ϖ_0 is shown in Table 13.4. The results in Table 13.4 show clearly that $\varpi_0 = 1$, called the *conservative case* [10], is a limiting situation from the numerical as well as the physical standpoint.

TABLE 13.4

Number of iterations required to obtain the solutions in Table 13.3

ϖ_0	Number of Iterations
0.1	7
0.2	8
0.3	10
0.4	12
0.5	14
0.6	17
0.7	21
0.8	28
0.9	43
1.0	10,587

Once the solutions ξ_i, $i = 1, 2, \ldots, 9$, of (13.40) have been obtained for a given value of ϖ_0, the question of their relationship to the corresponding solution $H(\mu)$ of (13.1) arises. One approach to answering this question is to construct a function by some interpolation procedure, using the values obtained from the solution of the arithmetic problem. This would ordinarily be done, since the nodes of the gaussian numerical integration rule are not the values of μ at which approximate solutions are usually desired; for example, we could employ linear interpolation (see Problem 6, Exercises 3) by making use of the fact that

$$H(0) = 1, \qquad 0 \leq \varpi_0 \leq 1 \qquad (13.46)$$

from (13.1) and extending the straight-line segment joining (μ_8, ξ_8) and (μ_9, ξ_9) to $\mu = 1$. This is equivalent to using the basis

$$\phi_0(\mu) = \begin{cases} 1 - \dfrac{\mu}{\mu_1}, & 0 \leq \mu \leq \mu_1, \\ 0, & \mu_1 \leq \mu \leq 1, \end{cases}$$

$$\begin{aligned} \phi_i(\mu) &= 0, & 0 \leq \mu \leq \mu_{i-1}, \\ &= \dfrac{\mu - \mu_{i-1}}{\mu_i - \mu_{i-1}}, & \mu_{i-1} \leq \mu \leq \mu_i, \\ &= \dfrac{\mu_{i+1} - \mu}{\mu_{i+1} - \mu_i}, & \mu_i \leq \mu \leq \mu_{i+1}, \\ &= 0, & \mu_{i+1} \leq \mu \leq 1, \end{aligned} \qquad (13.47)$$

where $i = 1, 2, \ldots, 8$,

$$\phi_9(\mu) = \begin{cases} 0, & 0 \leq \mu \leq \mu_8, \\ \dfrac{\mu - \mu_8}{\mu_9 - \mu_8}, & \mu_8 \leq \mu \leq 1, \end{cases}$$

for a 10-dimensional subspace of $C[0, 1]$. The resulting function is

$$\tilde{H}(\mu) = \phi_0(\mu) + \sum_{i=1}^{9} \xi_i \, \phi_i(\mu), \qquad 0 \leq \mu \leq 1. \qquad (13.48)$$

Since the integral transforms

$$\Phi_i(\mu) = \int_0^1 \frac{\mu}{\mu + \mu'} \, \phi_i(\mu') \, d\mu', \qquad 0 \leq \mu \leq 1, \quad i = 0, 1, \ldots, 9, \qquad (13.49)$$

can be calculated explicitly, we could set

$$H_0(\mu) = \tilde{H}(\mu), \qquad 0 \leq \mu \leq 1,$$

and obtain $H_1(\mu)$ by the iterative formula (13.6). On the basis of this information we can check the validity of the hypotheses of Theorem 12.2. If they are satisfied, (12.11) will give a rigorous error bound for $H_1(\mu)$ constructed by the numerical process described.

Another method of interpolation makes use of the approximate integral transform [see (13.33)],

$$\int_0^1 \frac{H(\mu')}{\mu + \mu'} \, d\mu' \cong \sum_{j=1}^9 \frac{w_j \xi_j}{\mu + \mu_j}, \qquad 0 \le \mu \le 1, \tag{13.51}$$

using the values for ξ_i, $i = 1, 2, \ldots, 9$ found numerically. Substitution of (13.51) into (13.1) yields

$$H(\mu) \cong 1 + \tfrac{1}{2}\varpi_0 H(\mu) \sum_{j=1}^9 \frac{w_j \xi_j}{\mu + \mu_j}, \qquad 0 \le \mu \le 1,$$

or the approximation to $H(\mu)$

$$\hat{H}(\mu) = \left(1 - \tfrac{1}{2}\varpi_0 \mu \sum_{j=1}^9 \frac{w_j \xi_j}{\mu + \mu_j}\right)^{-1}, \qquad 0 \le \mu \le 1. \tag{13.53}$$

This is called the *natural* method of interpolation for an integral equation, since it makes direct use of a continuous approximation to the integral transform. For $\varpi_0 = 0.5$, the values obtained by using (13.53) agree very closely with the exact solution [16]. In principle, we could take $H_0(\mu) = \hat{H}(\mu)$ and employ the same error estimation procedure described for (13.48). The analytic difficulties, however, would be considerable.

The application of the contraction mapping principle in this section was based on some more general considerations useful in practice. For a continuous operator F the quantity

$$\|F\| = \theta(r), \qquad r \ge 0, \tag{13.54}$$

in the ball $\bar{U}(x_0, r)$ is a nonnegative, continuous, monotone increasing function of r.

Suppose that $\theta(0) < 1$ and R is the smallest positive number such that $\theta(R) = 1$. (If $\|F\| < 1$ in any ball of finite radius, we may set $R = \infty$.) It is also true that

$$r_0(r) = \frac{1}{1 - \theta(r)} \|F(x_0) - x_0\|, \qquad 0 \le r < R, \tag{13.55}$$

is a nonnegative, continuous, monotone increasing function of r. Inequality (12.10) implies that the hypotheses of Theorem 12.2 will be satisfied if the curve (13.55) intersects the straight line

$$r_0 = r \tag{13.56}$$

(see Figure 13.1). The intersection point with the smallest abcissa $r_0{}^E$ indicates that there is a fixed point x^* of F in $\bar{U}(x_0, r_0{}^E)$. The intersection point with the largest abcissa $r_0{}^U$ determines the ball $\bar{U}(x_0, r_0{}^U)$ in which Theorem 12.2 guarantees that x^* is unique. Of course, it is possible that there is no such intersection, in which case x^* is unique in X, if $R = \infty$.

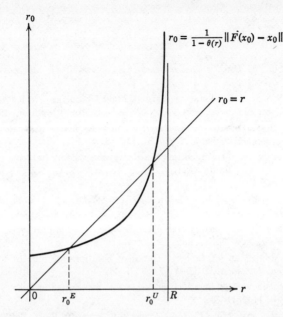

$$r_0 = \frac{1}{1 - \theta(r)} \| \vec{F}(x_0) - x_0 \|$$

$$r_0 = r$$

Figure 13.1 Graphical determination of $\bar{U}(x_0, r_0{}^E)$ and $\bar{U}(x_0, r^U{}_0)$.

EXERCISES 13

1. Investigate the use of the initial approximation
$$H_0(\mu) = 1 + c\mu, \qquad 0 \le \mu \le 1,$$
to the solution of (13.1). Try to choose the constant c to extend the range (13.27) of values of ϖ_0 for which this is a suitable initial approximation.

2. Construct an arithmetic version of (13.1) on the basis of *Simpson's rule*,
$$\int_0^1 f(s)\, ds \cong \tfrac{1}{6}[f(0) + 4f(\tfrac{1}{2}) + f(1)].$$

Find the range of ϖ_0 for which $x_0 = (1, 1, 1)$ is a suitable initial approximation. Solve the arithmetic problem for $\varpi_0 = 0.5$ and obtain approximate solutions of (13.1) by linear and natural interpolation. Discuss error estimation for each method.

3. Calculate the integral transforms (13.49). Use the result to obtain $H_1(\mu)$ for $\varpi_0 = 0.5$. Discuss error estimation for $H_1(\mu)$.

4. Investigate the applicability of the contraction mapping principle to the equivalent form of (13.1) given by
$$H(\mu) = \left(1 - \tfrac{1}{2}\varpi_0\mu \int_0^1 \frac{H(\mu')}{\mu + \mu'}\, d\mu'\right)^{-1}.$$

Construct an arithmetic model of this equation.

5. Use (13.31) to determine the maximum values of ϖ_0 for which $\|H - H_1\| \le 0.001$, $\|H - H_1\| \le 0.01$, and $\|H - H_1\| \le 0.1$ respectively.

REFERENCES AND NOTES

[1]   E. T. Bell. *The Development of Mathematics*. Second ed. McGraw-Hill, New York 1945.

[2]   Arthur Wouk. Direct Iteration, Existence and Uniqueness. In P. M. Anselone (ed.), *Nonlinear Integral Equations*. University of Wisconsin Press, Madison, 1964, 3–33.

[3]   S. Banach. *Sur les opérations dans les ensembles abstraits et leur applications aux equations integrales. Fund. Math.*, **3** (1922), 133–181.

[4]   R. Caccioppoli. Sugli elementi uniti delle transformazioni funzionali: un'osservazione sui problemi di valori ai limiti. *Atti Accad. Naz. Lincei Mem. Cl. Sci. Fis. Mat. Natur. Sez. I*, **6** (1931), 498–502.

[5]   Johannes Weissinger. *Zur Theorie und Anwendung des Iterationsverfahrens. Math. Nachr.*, **8** (1952), 193–212.

[6]   D. Greenspan. *Introductory Numerical Analysis of Elliptic Boundary Value Problems.* Harper and Row, New York, 1965.

[7]   C. M. Ablow and C. L. Perry. Iterative Solutions of the Dirichlet Problem for $\Delta u = u^2$. *J. Soc. Indust. Appl. Math.*, **7** (1959), 459–467.

[8]   S. I. Pohožaev. The Dirichlet Problem for the Equation $\Delta u = u^2$. *Soviet Math. Dokl.*, **1** (1960), 1143–1146.

[9]   W. E. Milne. *Numerical Calculus*. Princeton University Press, 1949.

[10]   S. Chandrasekhar. *Radiative Transfer*. Dover, New York, 1960.

[11]   V. A. Ambartsumian (ed.). *Theoretical Astrophysics*. Tr. by J. B. Sykes. Pergamon, New York, 1948.

[12]   D. W. N. Stibbs and R. E. Weir. On the H-functions for Isotropic Scattering. *Monthly Not. Royal Astron. Soc.*, **119** (1959), 512–525.

[13]   B. Noble. The Numerical Solution of Nonlinear Integral Equations and Related Topics. In P. M. Anselone (ed.), *Nonlinear Integral Equations*. University of Wisconsin Press, Madison, 1964, 215–318.

[14]   P. J. Davis and P. Rabinowitz. *Numerical Integration*. Blaisdell, Waltham, Massachusetts, 1967.

[15]   Table 13.2 was calculated with the aid of an LGP-30 computer at Lamar State College of Technology. Tables 13.3 and 13.4 were obtained by use of a CDC 3600 computer at the University of Wisconsin.

[16]   L. B. Rall. Numerical Integration and the Solution of Integral Equations by the Use of Riemann Sums. *SIAM Rev.*, **7** (1965), 55–64.

3

DIFFERENTIATION OF OPERATORS

The theory of nonlinear equations developed in the previous chapter depended only on the simple analytic notions of boundedness and continuity of the operator F involved. The computational procedures to be considered later require, in addition, that the operator be differentiable in some generalized sense. This chapter is concerned with the extension of some of the basic ideas of the differential calculus to operators in Banach spaces. As in the scalar case this calculus not only provides powerful tools for the study of nonlinear operators of importance in applications but is also interesting in its own right.

14. THE FIRST DERIVATIVE

There are several possible definitions for the derivative of an operator which maps one Banach space into another, each of which reduces to the customary definition in the case of scalar functions. The definition adopted here is the one due to Fréchet [1], which is simple in concept and easy to apply. Unless otherwise noted, the words *derivative* and *differential* will be used to denote the Fréchet derivative, and Fréchet differential respectively.

Definition 14.1. Suppose that P is an operator that maps a Banach space X into a Banach space Y. If a bounded linear operator L from X into Y exists such that

$$\lim_{\|\Delta x\| \to 0} \frac{\|P(x_0 + \Delta x) - P(x_0) - L\,\Delta x\|}{\|\Delta x\|} = 0, \qquad (14.1)$$

then P is said to be *differentiable at* x_0, and the bounded linear operator

$$P'(x_0) = L \qquad (14.2)$$

is called the *first derivative of P at* x_0.

Note that the limit (14.1) is supposed to hold independently of the manner in which Δx approaches 0. It follows at once from (14.1) that P is continuous at x_0 if it is differentiable at x_0. Furthermore, the *Fréchet differential*

$$\delta P(x_0, \Delta x) = P'(x_0) \, \Delta x \qquad (14.3)$$

is an arbitrarily close approximation to the difference $P(x_0 + \Delta x) - P(x_0)$, relative to $\|\Delta x\|$, for $\|\Delta x\|$ sufficiently small.

Certain operators may be differentiated by direct application of Definition 14.1; for example, if L is a bounded linear operator from X into Y, then

$$L'(x_0) = L \qquad (14.4)$$

for any $x_0 \in X$. If P and Q are operators from X into Y, their *sum* $P + Q$ is the operator defined by

$$(P + Q)(x) = P(x) + Q(x) \qquad (14.5)$$

for all $x \in X$. If P and Q are differentiable at $x = x_0$, then

$$(P + Q)'(x_0) = P'(x_0) + Q'(x_0); \qquad (14.6)$$

that is, *the derivative of the sum of two operators is the sum of their derivatives*.

If Q is an operator from a Banach space X into a Banach space Z, and P is an operator from Z into a Banach space Y, then their *product* (or *composition*) PQ is the operator from X into Y defined by

$$PQ(x) = P(Q(x)) \qquad (14.7)$$

for all $x \in X$. This definition is a generalization of (9.2), which was framed for the case $X = Y = Z$. It follows from Definition 14.1 that PQ will be differentiable at x_0 if Q is differentiable at x_0 and P is differentiable at the point $z_0 = Q(x_0)$ of Z, with

$$(PQ)'(x_0) = P'(Qx_0) \, Q'(x_0). \qquad (14.8)$$

This is the *chain rule* of scalar differential calculus, which may be expressed in operator terminology by saying that *the derivative of the product of two operators is the product of their derivatives*. (The ordinary rule for differentiating the multiplicative product of scalar functions will be shown later to be a consequence of the rule for differentiating bilinear operators.)

By combining (14.4) with (14.8), if L is a bounded linear operator from Z into Y, and Q is an operator from X into Z which is differentiable at $x = x_0$, then LQ is differentiable at x_0 and

$$(LQ)'(x_0) = LQ'(x_0). \tag{14.9}$$

Thus in the differential calculus in Banach spaces, bounded linear operators play roles similar to constant multipliers in ordinary calculus.

In practice, to differentiate a given nonlinear operator P we would attempt to write the difference $P(x_0 + \Delta x) - P(x_0)$ in the form

$$P(x_0 + \Delta x) - P(x_0) = L(x_0, \Delta x)\, \Delta x + \eta(x_0, \Delta x), \tag{14.10}$$

where $L(x_0, \Delta x)$ is a bounded linear operator for given x_0, Δx, with

$$\lim_{\|\Delta x\| \to 0} L(x_0, \Delta x) = L, \tag{14.11}$$

a bounded linear operator, and

$$\lim_{\|\Delta x\| \to 0} \frac{\|\eta(x_0, \Delta x)\|}{\|\Delta x\|} = 0. \tag{14.12}$$

If (14.11) and (14.12) hold, then

$$\lim_{\|\Delta x\| \to 0} L(x_0, \Delta x) = P'(x_0). \tag{14.13}$$

Also, if $L(x_0, \Delta x)$, considered to be an operator from X into $L(X, Y)$, the space of bounded linear operators from X into Y, is a continuous function of Δx in some ball $U(0, r)$ for $r > 0$, then

$$L(x_0, 0) = P'(x_0). \tag{14.14}$$

A formal description may be given for the process of differentiation. First we define a *mosaic* to be a given one- or two-dimensional array of symbols. For example, this page is a mosaic, as is (14.1), or

$$\begin{matrix} A & A & B & A & A & C & A & D \\ T & T & U & A & A & D & A & C. \\ D & A & C & A & A & B & A & A \end{matrix} \tag{14.15}$$

A mosaic consisting entirely of symbols written on a single line is called a *string*; thus the last word in this sentence is an example of a string. To distinguish a mosaic from its surroundings, it will sometimes be enclosed in parentheses, so that (14.15) may be written as

$$\begin{pmatrix} A & A & B & A & A & C & A & D \\ T & T & U & A & D & A & A & C \\ D & A & C & A & A & B & A & A \end{pmatrix}, \tag{14.16}$$

the parentheses not being considered to be part of the mosaic they enclose.

From a mosaic we may obtain a *formal operator* by removing a symbol from the mosaic, and inserting an appropriate *place holder* [ ] to indicate the position of the deleted symbol. Thus,

$$
\begin{pmatrix}
A & A & B & A & [\,] & C & A & D \\
T & T & U & A & A & D & A & C \\
D & A & C & A & A & B & A & A
\end{pmatrix}
\tag{14.17}
$$

is a formal operator obtained from the mosaic (14.16). A formal operator operates on a symbol to yield a mosaic in which the operand symbol occupies the position indicated by the place holder. Hence

$$
\begin{pmatrix}
A & A & B & A & [\,] & C & A & D \\
T & T & U & A & A & D & A & C \\
D & A & C & A & A & B & A & A
\end{pmatrix} A =
\begin{pmatrix}
A & A & B & A & A & C & A & D \\
T & T & U & A & A & D & A & C \\
D & A & C & A & A & B & A & A
\end{pmatrix},
\tag{14.18}
$$

the original mosaic (14.16) from which the operator (14.17) was obtained. Of course, (14.17) could operate on a different symbol, as

$$
\begin{pmatrix}
A & A & B & A & [\,] & C & A & D \\
T & T & U & A & A & D & A & C \\
D & A & C & A & A & B & A & A
\end{pmatrix} Z =
\begin{pmatrix}
A & A & B & A & Z & C & A & D \\
T & T & U & A & A & D & A & C \\
D & A & C & A & A & B & A & A
\end{pmatrix},
\tag{14.19}
$$

to yield a different mosaic.

If the result of applying a formal operator to a symbol representing an element of some set X (such as a Banach space) can be interpreted as representing an element of a set Y, then the formal operator is actually an operator from X into Y.

In terms of these concepts it is possible to define the following formal procedure for differentiation.

STEP 1. Write the difference $P(x_0 + \Delta x) - P(x_0)$ as a mosaic consisting of *terms* (elements of Y) separated by plus or minus signs.

STEP 2. Convert each term into a formal operator on Δx, considering Δx to be a single symbol, and regard the result to be the operator $L(x_0, \Delta x)$ on Δx as defined by (14.10), taking $\eta(x_0, \Delta x) = 0$.

STEP 3. Apply (14.13), or (14.14) in the case of continuity of $L(x_0, \Delta x)$ with respect to Δx, to obtain $P'(x_0)$, it being required that the limit satisfies Definition 14.1.

The second step replaces the operation of division by Δx which is called for by the famous *three step rule* of elementary calculus [2], which requires the assumption that $\Delta x \neq 0$.

To illustrate this process consider the operator P in $C[0, 1]$ defined by

$$P(x) = x(s) \int_0^1 \frac{s}{s + t} \, x(t) \, dt, \qquad 0 \leq s \leq 1, \qquad (14.20)$$

which is closely related to the operator F defined by (13.12). As the result of step 1,

$$P(x_0 + \Delta x) - P(x_0) = x_0(s) \int_0^1 \frac{s}{s + t} \Delta x(t) \, dt + \Delta x(s) \int_0^1 \frac{s}{s + t} \, x_0(t) \, dt$$

$$+ \Delta x(s) \int_0^1 \frac{s}{s + t} \Delta x(t) \, dt. \quad (14.21)$$

From step 2

$$L(x_0, \Delta x) = x_0(s) \int_0^1 \frac{s}{s + t} [\ ] \, dt + [\ ] \int_0^1 \frac{s}{s + t} \, x_0(t) \, dt$$

$$+ [\ ] \int_0^1 \frac{s}{s + t} \Delta x(t) \, dt. \quad (14.22)$$

For given $x_0(t)$ and $\Delta x(t)$, (14.22) does define a bounded linear operator in $C[0, 1]$, which as a linear operator is a continuous function of Δx. Note that there is more than one way to express the result of step 2; for example, the place holder and Δx can be interchanged in the last term of (14.22). By the continuity of $L(x_0, \Delta x)$ we may set $\Delta x = 0$ in step 3 to obtain

$$P'(x_0) = x_0(s) \int_0^1 \frac{s}{s + t} [\ ] \, dt + [\ ] \int_0^1 \frac{s}{s + t} \, x_0(t) \, dt. \qquad (14.23)$$

For $y = y(s)$ in $C[0, 1]$,

$$P'(x_0)y = x_0(s) \int_0^1 \frac{s}{s + t} \, y(t) \, dt + y(s) \int_0^1 \frac{s}{s + t} \, x_0(t) \, dt, \qquad 0 \leq s \leq 1. \qquad (14.24)$$

The linear operator defined by (14.23) is consequently the sum of a linear integral transform K_0 with kernel

$$K_0(s, t) = x_0(s) \frac{s}{s + t}, \qquad 0 \leq s, \ t \leq 1, \qquad (14.25)$$

and ordinary multiplication of the operand function $y(s)$ by the continuous function

$$K x_0 = \int_0^1 \frac{s}{s + t} \, x_0(t) \, dt, \qquad 0 \leq s \leq 1. \qquad (14.26)$$

Alternatively, we could use the formulation (14.10), with

$$L(x_0, \Delta x) = x_0(s) \int_0^1 \frac{s}{s+t} [\ ]\, dt + [\ ] \int_0^1 \frac{s}{s+t} x_0(t)\, dt \qquad (14.27)$$

and

$$\eta(x_0, \Delta x) = \Delta x(s) \int_0^1 \frac{s}{s+t} \Delta x(t)\, dt. \qquad (14.28)$$

Other examples of differentiation of operators are given in the next section.

EXERCISES 14

1. Prove (14.6).
2. Prove (14.8).
3. Find the derivative of the operator P in R_∞^2 defined by

$$P\begin{pmatrix} \xi \\ \eta \end{pmatrix} = \begin{pmatrix} \xi^2 - 2\xi + 3\xi\eta^4 - 1 \\ \xi(\xi + \eta)^2 \end{pmatrix}$$

at $x_0 = (\xi_0, \eta_0)$. *Hint.* $P'\begin{pmatrix} \xi_0 \\ \eta_0 \end{pmatrix}$ must have the form of a 2×2 matrix. Why?

15. EXAMPLES OF DERIVATIVES

In elementary calculus a number of functions occur with sufficient frequency to make it worthwhile to tabulate their derivatives. Similarly, certain types of operators between specific Banach spaces occur often in computational problems. Explicit representations for the derivatives of a number of important operators will be derived in this section for future reference.

In actual computation, using digital computers, we work with the Banach spaces $X = R_p^n$, $Y = R_p^m$, for some value of p in the range $1 \le p \le \infty$ (usually 1, 2, or ∞). An operator P such that $P(x) = y$, $x \in X$, $y \in Y$, has the representation

$$\eta_i = P_i(\xi_1, \xi_2, \ldots, \xi_n), \qquad i = 1, 2, \ldots, m, \qquad (15.1)$$

for $x = (\xi_1, \xi_2, \ldots, \xi_n)$, $y = (\eta_1, \eta_2, \ldots, \eta_m)$. Suppose that the partial derivatives

$$\frac{\partial P_i}{\partial \xi_j}\bigg|_{x=x_0} = \frac{\partial P_i(\xi_1^{(0)}, \xi_2^{(0)}, \ldots, \xi_n^{(0)})}{\partial \xi_j}, \qquad i = 1, 2, \ldots, m, \quad j = 1, 2, \ldots, n,$$

$$(15.2)$$

exist and are continuous at the point $x_0 = (\xi_1^{(0)}, \xi_2^{(0)}, \ldots, \xi_n^{(0)}$. Then the derivative of P is represented by the *Jacobian matrix*

$$P'(x_0) = \begin{vmatrix} \dfrac{\partial P_1}{\partial \xi_1} & \dfrac{\partial P_1}{\partial \xi_2} & \cdots & \dfrac{\partial P_1}{\partial \xi_n} \\[2mm] \dfrac{\partial P_2}{\partial \xi_1} & \dfrac{\partial P_2}{\partial \xi_2} & \cdots & \dfrac{\partial P_2}{\partial \xi_n} \\[2mm] \cdot \\ \cdot \\ \cdot \\ \dfrac{\partial P_m}{\partial \xi_1} & \dfrac{\partial P_m}{\partial \xi_2} & \cdots & \dfrac{\partial P_m}{\partial \xi_n} \end{vmatrix}_{x=x_0} . \tag{15.3}$$

To see this, note that the matrix $P'(x_0) = (\partial P_i/\partial \xi_j)_{x=x_0}$ is a bounded linear operator from R_p^n into R_p^m, since it is an $m \times n$ matrix with constant coefficients. By the definition of the partial derivative of a real function of several variables Definition 14.1 is satisfied for $p = \infty$ and thus for all values of p by the equivalence of norms for finite-dimensional spaces.

If $m = n = 10$, for instance, then (15.3) would contain 100 entries and could be tedious to calculate. Fortunately it is possible to program computers to perform the differentiation of a wide class of functions analytically and to prepare programs automatically for the evaluation of the required derivatives [3]. This capability of computing machines makes possible the effective treatment of a large number of nonlinear problems which were previously inaccessible for all practical purposes. These computer programs for analytic differentiation make use of the fact that most functions which occur in practice can be differentiated by applying simple rules such as (14.6) and (14.8) and using the formulas for the derivatives of a few standard functions.

As an example of application of (15.3), the derivative of a real function

$$f = f(\xi, \eta), \tag{15.4}$$

considered to be a mapping from R_p^2 into R, is the 1×2 matrix

$$f'(x_0) = \left(\dfrac{\partial f(\xi_0, \eta_0)}{\partial \xi} \quad \dfrac{\partial f(\xi_0, \eta_0)}{\partial \eta} \right), \tag{15.5}$$

where $x_0 = (\xi_0, \eta_0)$. For the *increment* $dx = (d\xi, d\eta)$, we have the *differential*

$$df = \delta f(x_0, dx) = f'(x_0)\, dx = \dfrac{\partial f(\xi_0, \eta_0)}{\delta \xi}\, d\xi + \dfrac{\partial f(\xi_0, \eta_0)}{\partial \eta}\, d\eta \tag{15.6}$$

[see (14.3)] as a linear approximation to $f(x_0 + dx) - f(x_0)$. In calculus (15.6) is customarily called the *total differential* of $f(\xi, \eta)$.

The function (15.4) is a particular example of a *functional*, that is, an operator which maps a Banach space X into its scalar field Λ. The derivative of a functional f at $x_0 \in X$ must, by Definition 14.1, be a bounded linear functional on X. The set of all bounded linear functionals on X forms a Banach space (recall Theorem 8.4), which is denoted by X^* and called the *conjugate space* to X. If X is a Hilbert space, then $X = X^*$ by the theorem of F. Riesz cited in Section 8. Thus if f is a functional on a Hilbert space H which is differentiable at $x_0 \in H$, then

$$f'(x_0) = y_0, \tag{15.7}$$

where $y_0 \in H$ is the *kernel* of the linear functional

$$f'(x_0)h = \langle h, y_0 \rangle \tag{15.8}$$

for all $h \in H$.

A mapping $x = x(\lambda)$ into a Banach space X from its scalar field Λ is called an *abstract function*. A linear abstract function y is represented uniquely by an element $y \in X$, where

$$y\lambda = \lambda y, \qquad \lambda \in \Lambda; \tag{15.9}$$

that is, the element resulting from operation on the scalar λ by the operator y is simply obtained by scalar multiplication of the element y by λ. If $x(\lambda)$ is an abstract function that is differentiable at λ_0, then

$$x'(\lambda_0) = y_0, \tag{15.10}$$

where y_0 is some fixed element of X.

Some Banach spaces, such as $C[0, 1]$, have the property that the *product* xy of two elements of the space is defined and is an element of the space. Such a Banach space X is called a *Banach algebra* if

$$\|xy\| \leq \|x\| \cdot \|y\| \tag{15.11}$$

for $x, y \in X$. A Banach algebra is said to be *commutative* if

$$xy = yx \tag{15.12}$$

for arbitrary $x, y \in X$. If (15.12) fails to hold for at least one pair x, y, then the algebra is said to be *noncommutative*. The spaces $C[0, 1]$ and $L_2[0, 2\pi]$ are examples of commutative Banach algebras. The space $L(R_p{}^2, R_p{}^2)$ of 2×2 real matrices is a noncommutative Banach algebra.

In a Banach algebra X the linear operator defined by multiplication by a fixed element x is denoted by xI or $x[\ ]$. If the algebra is noncommutative, then we must distinguish between xI or $x[\ \ ]$ and Ix or $[\ \ ]x$, where

$$(xI)y = (x[\ \ ])y = xy,$$
$$(Ix)y = ([\ \ ]x)y = yx,$$
(15.13)

for all $y \in X$. These concepts and notations are introduced to facilitate the calculation of the derivatives to follow. It is also convenient to introduce for a function $f(\xi_1, \xi_2, \ldots, \xi_n)$ of several variables the notation

$$f_j'(\xi_1, \xi_2, \ldots, \xi_n) = \frac{\partial f(\xi_1, \xi_2, \ldots, \xi_n)}{\partial \xi_j}, \qquad 1 \leq j \leq n, \quad (15.14)$$

for the indicated partial derivative.

In $C[0, 1]$ the *Hammerstein integral operator* H is defined by

$$H(x) = \int_0^1 K(s, t)\, g(t, x(t))\, dt, \qquad (15.15)$$

for $x \in C[0, 1]$, where $K(s, t)$ is the kernel of a linear integral operator in $C[0, 1]$ and $g(t, u)$ is continuous for $0 \leq t \leq 1$, $-\infty < u < +\infty$. The operator H may be written as

$$H = KG, \qquad (15.16)$$

where K is the linear integral operator with kernel $K(s, t)$ and G is the *Nemyckiĭ operator* for which

$$G(x) = g(s, x(s)), \qquad 0 \leq s \leq 1, \qquad (15.17)$$

for $x \in C[0, 1]$. It follows that

$$G'(x_0) = g_2'(s, x_0(s))I, \qquad (15.18)$$

a linear operator in $C[0, 1]$ and, from (14.9),

$$H'(x_0) = KG'(x_0). \qquad (15.19)$$

Thus $H'(x_0)$ is a linear integral operator with kernel

$$H'(x_0) = K(s, t)g_2'(t, x_0(t)), \qquad 0 \leq s, t \leq 1, \qquad (15.20)$$

which may be verified by differentiating (15.15) directly. Thus for $y \in C[0, 1]$,

$$H'(x_0)y = \int_0^1 K(s, t)g_2'(t, x_0(t))\, y(t)\, dt. \qquad (15.21)$$

As another example, consider the partial differential operator

$$P(u) = \Delta u - u^2 \qquad (15.22)$$

from $C^2(S)$ into $C(S)$, the space of all continuous functions on the square $0 \leq \xi, \eta \leq 1$, which was considered in Section 11. We find that

$$P'(u_0) = \Delta - 2u_0 I, \qquad (15.23)$$

since the Laplace operator Δ defined by (11.19) is a linear operator. Thus for $v \in C^2(S)$,

$$P'(u_0)v = \Delta v(\xi, \eta) - 2u_0(\xi, \eta) v(\xi, \eta). \tag{15.24}$$

For any Banach space X the space $L(X, X)$—which will be denoted by $L(X)$ for brevity—of bounded linear operators in X will be a Banach algebra, generally noncommutative. For the operator

$$P(L) = L^3 \tag{15.25}$$

in $L(X)$ we have, since L_0 and ΔL do not necessarily commute,

$$\begin{aligned}
P(L_0 + \Delta L) - P(L_0) &= (L_0 + \Delta L)^3 - L_0{}^3 \\
&= L_0 \Delta L L_0 + L_0{}^2 \Delta L + L_0 \Delta L^2 + \Delta L L_0{}^2 \\
&\quad + \Delta L^2 L_0 + \Delta L L_0 \Delta L + \Delta L^3.
\end{aligned} \tag{15.26}$$

Equation 15.26 may be expressed in the form (14.10) as

$$\begin{aligned}
P(L_0 + \Delta L) - P(L_0) &= (L_0[\ \ ]L_0 + L_0{}^2[\ \ ] + [\ \ ]L_0{}^2) \Delta L + L_0 \Delta L^2 \\
&\quad + \Delta L^2 L_0 + \Delta L L_0 \Delta L + \Delta L^3,
\end{aligned} \tag{15.27}$$

so that

$$P'(L_0) = L_0[\ \ ]L_0 + L_0{}^2[\ \ ] + [\ \ ]L_0{}^2. \tag{15.28}$$

The *inversion operator* defined by

$$P(L) = L^{-1} \tag{15.29}$$

for all linear operators L which have inverses may also be differentiated at any L_0 at which it is defined. By Theorem 10.1, if $L_0{}^{-1}$ exists, then $(L_0 + \Delta L)^{-1}$ exists for all ΔL in the ball $U(0, 1/\|L_0{}^{-1}\|)$ and, by (10.2),

$$(L_0 + \Delta L)^{-1} = \sum_{n=0}^{\infty} (-L_0{}^{-1} \Delta L)^n L_0{}^{-1}. \tag{15.30}$$

Consequently,

$$\begin{aligned}
P(L_0 + \Delta L) - P(L_0) &= \sum_{n=1}^{\infty} (-L_0{}^{-1} \Delta L)^n L_0{}^{-1} \\
&= -L_0{}^{-1} \Delta L L_0{}^{-1} + \sum_{n=2}^{\infty} (-L_0{}^{-1} \Delta L)^n L_0{}^{-1}.
\end{aligned} \tag{15.31}$$

It follows immediately from (15.31) that

$$P'(L_0) = -L_0{}^{-1}[\ \ ]L_0{}^{-1}. \tag{15.32}$$

When calculating the derivative of an operator from a space X into a space Y, it should be kept in mind that we are seeking a bounded linear operator from X into Y, that is, an element of $L(X, Y)$. If the representation of elements of $L(X, Y)$ is known, then this determines the representation of the derivative.

EXERCISES 15

1. If f is a functional on $R_p{}^n$, $1 \leq p \leq \infty$, prove that $f'(x_0)$ may be represented by $y_0 = (\eta_1^{(0)}, \eta_2^{(0)}, \ldots, \eta_n^{(0)})$, with

$$f'(x_0)x = \sum_{i=1}^{n} \xi_i \eta_i^{(0)},$$

for $x = (\xi_1, \xi_2, \ldots, \xi_n)$ in $R_p{}^n$.

2. Calculate the derivative of the *Urysohn integral operator*

$$U(x) = \int_0^1 f(s, t, x(t)) \, dt$$

in $C[0, 1]$ at $x_0 = x_0(s)$.

3. Calculate the derivative of the *Riccati differential operator*

$$R(y) = \frac{dy}{ds} + p(s)y^2 + q(s)y + r(s)$$

from $C'[0, T]$ into $C[0, T]$ at $y_0 = y_0(s)$ in $C'[0, T]$.

4. Calculate the derivative of the integrodifferential operator

$$P(x) = \int_0^1 f(s, t, x(t), x'(t)) \, dt, \qquad x'(s) = \frac{dx}{ds},$$

in $C'[0, 1]$ at $x_0 = x_0(s)$.

5. Calculate the derivative of the *hyperbolic* partial differential operator

$$P(x) = \frac{\partial^2 x(\xi, \eta)}{\partial \xi \, \partial \eta} - f(\xi, \eta, x(\xi, \eta))$$

from the space $C^2(R^2)$ of twice differentiable functions on R^2 into $C(R^2)$, the space of continuous functions on R^2, at $x_0 = x_0(\xi, \eta)$.

16. PARTIAL DERIVATIVES AND DIFFERENTIATION IN PRODUCT SPACES

In many applications we are led to consider operators which map the product space $X = X_1 \times X_2$ of two Banach spaces X_1, X_2 into a third Banach space X_3 of the form

$$x_3 = P(x_1, x_2). \tag{16.1}$$

For $x_2 = x_2^{(0)}$ fixed, (16.1) generates an operator P_1 defined by

$$P_1(x_1) = P(x_1, x_2^{(0)}) \tag{16.2}$$

which maps X_1 into X_3. If P_1 is differentiable at $x_1 = x_1^{(0)}$, then $P_1'(x_1^{(0)})$ is called the *partial derivative of P at $x_0 = (x_1^{(0)}, x_2^{(0)})$ with respect to x_1*, and we write

$$P_1'(x_1^{(0)}) = \frac{\partial P(x_1^{(0)}, x_2^{(0)})}{\partial x_1} \tag{16.3}$$

or

$$P_1'(x_1^{(0)}) = P_1'(x_1^{(0)}, x_2^{(0)}). \tag{16.4}$$

The abbreviated notation $\partial P/\partial x_1$ will also be used when the point of evaluation is understood. By definition, this partial derivative will be a linear operator from X_1 into X_3. Similarly, for $x_1 = x_1^{(0)}$ fixed,

$$P_2(x_2) = P(x_1^{(0)}, x_2) \tag{16.5}$$

is a mapping from X_2 into X_3, which, if differentiable at $x_2 = x_2^{(0)}$, leads to the definition of the partial derivative

$$P_2'(x_2^{(0)}) = \frac{\partial P(x_1^{(0)}, x_2^{(0)})}{\partial x_2} = P_2'(x_1^{(0)}, x_2^{(0)}) \tag{16.6}$$

of P with respect to x_2 at $x_0 = (x_1^{(0)}, x_2^{(0)})$.

In terms of the partial derivatives (16.3) and (16.6), the derivative $P'(x_0)$ of P at $x_0 = (x_1^{(0)}, x_2^{(0)})$ may be represented in the form of a 1×2 matrix of linear operators,

$$P'(x_0) = \left(\frac{\partial P}{\partial x_1} \quad \frac{\partial P}{\partial x_2}\right)_{x=x_0}, \tag{16.7}$$

which is a linear operator from $X = X_1 \times X_2$ into X_3 of the same type as considered in (7.14), Section 7. That is, for $y = (y_1, y_2)$ in X,

$$P'(x_0)y = \left(\frac{\partial P}{\partial x_1} \quad \frac{\partial P}{\partial x_2}\right)\binom{y_1}{y_2} = \frac{\partial P}{\partial x_1} y_1 + \frac{\partial P}{\partial x_2} y_2 \tag{16.8}$$

will be an element of X_3.

As an example, consider the mapping of $X = C[0, 1] \times C[0, 1]$ into $C[0, 1]$ by the operator P defined by

$$P(x, y) = x(s)\, y(s), \qquad 0 \le s \le 1. \tag{16.9}$$

Here

$$\frac{\partial P}{\partial x} = yI, \qquad \frac{\partial P}{\partial y} = xI, \tag{16.10}$$

and consequently,

$$P'(x_0, y_0) = (y_0 I \quad x_0 I). \tag{16.11}$$

The foregoing observations may be generalized at once to an operator P which maps a product space

$$X = \prod_{i=1}^{n} X_i \tag{16.12}$$

into a product space

$$Y = \prod_{i=1}^{m} Y_i \tag{16.13}$$

by means of the relationships

$$y_i = P_i(x_1, x_2, \ldots, x_n), \qquad i = 1, 2, \ldots, m. \tag{16.14}$$

For $x_0 = (x_1^{(0)}, x_2^{(0)}, \ldots, x_n^{(0)})$

$$\frac{\partial P_i(x_0)}{\partial x_j} = \frac{\partial P_i(x_1^{(0)}, x_2^{(0)}, \ldots, x_n^{(0)})}{\partial x_j}, \qquad i = 1, 2, \ldots, m, \qquad j = 1, 2, \ldots, n, \tag{16.15}$$

may be defined as above and will be a linear operator from the space X_j into the space Y_i. In this case $P'(x_0)$ can be represented as a matrix of linear operators of the same form a the ordinary Jacobian matrix; that is,

$$P'(x_0) = \begin{pmatrix} \dfrac{\partial P_1}{\partial x_1} & \dfrac{\partial P_1}{\partial x_2} & \cdots & \dfrac{\partial P_1}{\partial x_n} \\[2ex] \dfrac{\partial P_2}{\partial x_1} & \dfrac{\partial P_2}{\partial x_2} & \cdots & \dfrac{\partial P_2}{\partial x_n} \\[1ex] \cdot & & & \\ \cdot & & & \\ \cdot & & & \\ \dfrac{\partial P_m}{\partial x_1} & \dfrac{\partial P_m}{\partial x_2} & \cdots & \dfrac{\partial P_m}{\partial x_n} \end{pmatrix}_{x=x_0} \tag{16.16}$$

For example, for $x = (y, \xi)$ in $X = C'[0, T] \times R$, the operator P defined by $P = (P_1(x), P_2(x))$, with

$$P_1(x) = \frac{dy}{ds} - f(s, y(s)), \tag{16.17}$$

$$P_2(x) = \delta_0 y - \xi,$$

maps X into $C[0, T] \times R$. The functional δ_0 is the *evaluation functional* şuch that

$$\delta_0 y = y(0), \tag{16.18}$$

as defined in (2.6), Section 2. From (16.16) and (16.17), at $x_0 = (y_0, \xi_0)$,

$$P'(x_0) = \begin{pmatrix} \dfrac{d}{ds} - f_2'(s, y_0(s))I & 0 \\[2ex] \delta_0 & -1 \end{pmatrix}. \tag{16.19}$$

This matrix has the structure

$$
\begin{pmatrix}
\text{Linear operator} & \text{Abstract linear function} \\
C'[0, T] \text{ to } C[0, T] & R \text{ to } C[0, T] \\
& \\
\text{Linear functional} & \text{Constant multiplier} \\
C'[0, T] \text{ to } R & R \text{ to } R
\end{pmatrix},
$$

as the theory requires.

EXERCISES 16

1. The system of differential equations describing a chemical reaction

$$\frac{dx_1}{ds} + a_1 x_1 x_2 + a_2 x_3 + a_3 x_4 = y_1(s),$$

$$\frac{dx_2}{ds} + a_1 x_1 x_2 + a_2 x_3 = y_2(s),$$

$$\frac{dx_3}{ds} - a_1 x_1 x_2 + (a_2 + a_4)x_3 = y_3(s),$$

$$\frac{dx_4}{ds} - a_4 x_3 + a_3 x_4 = y_4(s),$$

where the a_i, $i = 1, 2, 3, 4$, are constants, is of the form $P(x) = y$, where P is an operator from $X = \{C[0', T]\}^4$ into $Y = \{C[0, T]\}^4$. Find $P'(x_0)$ at

$$x_0 = (x_1^{(0)}(s), x_2^{(0)}(s), x_3^{(0)}(s), x_4^{(0)}(s)).$$

2. If A is a bounded linear operator in a Hilbert space H with the scalar field C, an operator from $X = H \times C$ into itself is defined by $P(x) = (P_1(x), P_2(x))$, with

$$P_1(x) = Ah - \lambda h,$$

$$P_2(x) = \langle h, h \rangle - 1,$$

where $x = (h, \lambda)$, $h \in H$, $\lambda \in C$. Find $P'(x_0)$ for $x_0 = (h_0, \lambda_0)$. Compare the structure of the resulting matrix with (16.19).

17. MULTILINEAR OPERATORS AND ABSTRACT POLYNOMIALS

To give a precise definition of second and higher derivatives of nonlinear operators which map one Banach space into another, it is necessary to introduce some spaces and linear operators associated with the given spaces. It was shown in Section 8 (Theorem 8.4) that if X and Y are two given

Banach spaces the space $L(X, Y)$ of all bounded linear operators from X into Y is likewise a Banach space for the norm

$$\|L\| = \sup_{\|x\|=1} \|Lx\| \tag{17.1}$$

determined by the bound of $L \in L(X, Y)$ as a linear operator from X into Y. Since $L(X,Y)$ is itself a Banach space, the space $L(X, L(X, Y))$ of all bounded linear operators from X into $L(X, Y)$ will be another Banach space, and so on. These considerations give rise to the following definitions.

Definition 17.1. For $k = 1, 2, \ldots, L(X^{k+1}, Y)$ denotes the space of bounded linear operators from X into the space $L(X^k, Y)$, where $L(X^1, Y) = L(X, Y)$. If $X = Y$, then $L(X, X)$ is denoted by $L(X)$ and $L(X^k, X)$ by $L(X^k)$, $k = 1, 2, \ldots$. The spaces $L(X^k, Y)$, $k = 1, 2, \ldots$, are called the spaces *associated with* X, Y.

If X and Y are finite-dimensional, it is fairly easy to represent the spaces associated with them and the elements of the associated spaces; for example, if $X = R^n$, $Y = R^m$, the space $L(X, Y)$ is composed of $m \times n$ matrices

$$L = (\lambda_{ij}), \qquad i = 1, 2, \ldots, m, \quad j = 1, 2, \ldots, n, \tag{17.2}$$

with

$$Lx = \left(\sum_{j=1}^{n} \lambda_{ij} \xi_j \right), \qquad i = 1, 2, \ldots, m \tag{17.3}$$

being an element of Y for $x = (\xi_1, \xi_2, \ldots, \xi_n) \in X$. An element B of $L(X^2, Y)$ is represented by the $m \times n \times n$ array

$$B = (b_{ijk}), \qquad i = 1, 2, \ldots, m \cdot j, k = 1, 2, \ldots, n, \tag{17.4}$$

with

$$Bx = \left(\sum_{k=1}^{n} b_{ijk} \xi_k \right), \qquad i = 1, 2, \ldots, m, \quad j = 1, 2, \ldots, n, \tag{17.5}$$

being an $m \times n$ matrix. In general, an element G of $L(X^k, Y)$ would be represented by an array

$$G = (g_{ij_1 j_2 \cdots j_k}), \qquad i = 1, 2, \ldots, m, \quad j_1, j_2, \ldots, j_k = 1, 2, \ldots, n, \tag{17.6}$$

with

$$Gx = \left(\sum_{j_k=1}^{n} g_{ij_1 j_2 \cdots j_{k-1} j_k} \xi_{j_k} \right), \qquad i = 1, 2, \ldots, m,$$
$$j_1, j_2, \ldots, j_{k-1} = 1, 2, \ldots, n, \tag{17.7}$$

being an element of $L(X^{k-1}, Y)$, for $k \geq 2$.

Definition 17.2. An element of $L(X^k, Y)$ is called a *k-linear operator from* X *into* Y. It is said to be *in* X if $X = Y$. For $k = 1$ the elements of $L(X,Y)$

are referred to simply as *linear operators*. For $k > 1$, the elements of $L(X^k, Y)$ are called *multilinear operators* in general and, specifically, *bilinear operators* for $k = 2$ and *trilinear operators* for $k = 3$.

Bilinear operators were introduced previously in Section 7 [see (7.22) and Problem 3, Exercises 7]. A bilinear operator B defines a mapping of $X^2 = X \times X$ into Y by means of the relationship

$$y = B(x_1, x_2) = (Bx_1)x_2, \qquad x_1, x_2 \in X. \tag{17.8}$$

Adopting the convention that a multilinear operator acts on the adjacent point first,

$$Bx_1x_2 = (Bx_1)x_2 \tag{17.9}$$

is written for simplicity. In general, the order of the operations is significant; however, if

$$Bx_1x_2 = Bx_2x_1 \tag{17.10}$$

for all $x_1, x_2 \in X$, then B is said to be *symmetric*.

Since a bilinear operator B is a bounded linear operator from X into $L(X, Y)$,

$$\|Bx_1\| \leq \|B\| \cdot \|x_1\|, \tag{17.11}$$

and, thus,

$$\|Bx_1x_2\| \leq \|B\| \cdot \|x_1\| \cdot \|x_2\|, \tag{17.12}$$

for all $x_1, x_2 \in X$. Corresponding to each bilinear operator B is its *permutation* B^* defined by

$$B^*x_1x_2 = Bx_2x_1 \tag{17.13}$$

for all $x_1, x_2 \in X$, and its *mean* $\bar{B}$ given by

$$\bar{B} = \tfrac{1}{2}(B + B^*). \tag{17.14}$$

Obviously $\bar{B}$ is a symmetric bilinear operator.

For the finite-dimensional bilinear operator B given by (17.4), the operators

$$B^* = (b^*_{ijk}), \qquad \bar{B} = (\bar{b}_{ijk}), \qquad i = 1, 2, \ldots, m, \quad j, k = 1, 2, \ldots, n,$$

$$\tag{17.15}$$

defined by (17.13) and (17.14) respectively, have the components

$$b^*_{ijk} = b_{ikj} \tag{17.16}$$

and

$$\bar{b}_{ijk} = \tfrac{1}{2}(b_{ijk} + b_{ikj}), \qquad i = 1, 2, \ldots, m, \quad j, k = 1, 2, \ldots, n. \tag{17.17}$$

As a mapping from $X^2 = X \times X$ into Y, the operator

$$B(x_1, x_2) = Bx_1x_2, \tag{17.18}$$

where B is a bilinear operator from X into Y, has the derivative

$$B'(x_1^{(0)}, x_2^{(0)}) = (B^*x_2^{(0)} \quad Bx_1^{(0)}) \tag{17.19}$$

at $(x_1^{(0)}, x_2^{(0)})$, as may be verified directly from Definition 14.1. Hence for $X = Y = C[0, 1]$, tne bilinear operator defined by

$$B(x, y) = Bxy = x(s)\,y(s), \quad 0 \leq s \leq 1, \tag{17.20}$$

has the derivative

$$B'(x_0, y_0) = (y_0I \quad x_0I) \tag{17.21}$$

at (x_0, y_0), as found previously [see (16.11)]. If $x(s)$ and $y(s)$ are differentiable real functions, their derivatives at $s = s_0$ are $x'(s_0)$, $y'(s_0)$ respectively. As a mapping from the subset $[0, 1]$ of R into R, the function

$$f(s_0) = B(x(s_0), y(s_0)) = x(s_0)\,y(s_0) \tag{17.22}$$

may be differentiated by using the chain rule (14.8) and (17.21) to obtain

$$f'(s_0) = (y(s_0)I \quad x(s_0)I)\begin{pmatrix} x'(s_0) \\ y'(s_0) \end{pmatrix} = y(s_0)\,x'(s_0) + x(s_0)\,y'(s_0), \tag{17.23}$$

the ordinary rule for differentiating the multiplicative product of two functions. Of course, (17.23) could be obtained directly on the basis of Definition 14.1. This example shows the importance of distinguishing between the function $x(s)$ as an element of $C[0, 1]$ and the *evaluation* $x(s_0)$ of $x(s)$ at $s = s_0$ [4]. In the latter context x is actually an operator from R into R.

By Definition 17.2, for a given trilinear operator T from X into Y, Tx_1 is a bilinear operator from X into Y for $x_1 \in X$,

$$Tx_1x_2 = (Tx_1)x_2 \tag{17.24}$$

is a linear operator from X into Y for $x_1, x_2 \in X$, and

$$Tx_1x_2x_3 = ((Tx_1)x_2)x_3 \tag{17.25}$$

is an element of Y for $x_1, x_2, x_3 \in X$. Thus a trilinear operator T defines a mapping

$$T(x_1, x_2, x_3) = Tx_1x_2x_3 \tag{17.26}$$

from $X^3 = X \times X \times X$ into Y. In general, the points

$$
\begin{aligned}
&Tx_1x_2x_3, \quad Tx_1x_3x_2, \quad Tx_2x_1x_3, \\
&Tx_2x_3x_1, \quad Tx_3x_1x_2, \quad Tx_3x_2x_1,
\end{aligned}
\tag{17.27}
$$

are different, so that actually six trilinear operators and mappings (17.26) are associated with a given trilinear operator T, including T itself. These operators correspond to the six permutations (i_1, i_2, i_3) of the set of integers $(1, 2, 3)$.

One may define the trilinear operator $T_{i_1 i_2 i_3}$ by

$$T_{i_1 i_2 i_3} x_1 x_2 x_3 = Tx_{i_1} x_{i_2} x_{i_3} \tag{17.28}$$

to be the corresponding *permutation* of T. Obviously

$$T_{123} = T. \tag{17.29}$$

If all of the points (17.27) are identical for each given choice of x_1, x_2, $x_3 \in X$ or, equivalently, if all of the permutations of T are equal, then T is said to be *symmetric*. The *mean* $\bar{T}$ of T is the symmetric trilinear operator

$$\bar{T} = \tfrac{1}{6} \sum_{\pi(1,2,3)} T_{i_1 i_2 i_3}, \tag{17.30}$$

the sum being taken over all permutations $\pi(1, 2, 3)$ of the integers $(1, 2, 3)$.

In the same way a k-linear operator K defines a mapping from X^k into Y by means of the relationship

$$y = K(x_1, x_2, \ldots, x_k) = (\cdots((Kx_1)x_2)x_3 \cdots)x_k = Kx_1 x_2 \cdots x_k. \tag{17.31}$$

The operator K is called *symmetric* if

$$Kx_1 x_2 \cdots x_k = Kx_{i_1} x_{i_2} \cdots x_{i_k} \tag{17.32}$$

for every $(x_1, x_2, \ldots, x_k) \in X^k$ and all permutations $(i_1, i_2, \ldots, i_k)$ of the integers $(1, 2, \ldots, k)$. There are, in general, $k!$ permutations of K (including K itself), and the mean $\bar{K}$ of K is the symmetric k-linear operator formed by taking the average of all of the permutations of K. Actually $\bar{K}$ may be obtained by averaging just the distinct permutations of K, since some may be equal.

A k-linear operator, considered as a linear operator from X into the space of $(k - 1)$-linear operators from X into Y, generally will not have an inverse if $k \geq 2$. A typical example would be the bilinear operator

$$B = \begin{pmatrix} b_{111} & b_{112} & b_{121} & b_{122} \\ b_{211} & b_{212} & b_{221} & b_{222} \end{pmatrix} \tag{17.33}$$

from R^2 into R^2 [see Problem 3, Exercises 7, for an explanation of the notation (17.33)], which is a linear operator from the two-dimensional space R^2 into a subspace of the four-dimensional space of 2×2 matrices, the space of linear operators from R^2 into itself. However, such operators can be *nonsingular*, in the sense that their null spaces consist only of the origin 0 of X. This concept was introduced in Section 7, in particular by (7.26) and the subsequent remarks.

The null space of a symmetric multilinear operator K is characterized by the following theorem, which is a generalization of a theorem given in [5].

Theorem 17.1. If K is a symmetric $k -$ linear operator, then

$$N(K) = \bigcap_{(x_1, x_2, \ldots, x_{k-1}) \in X^{k-1}} N(Kx_1 x_2 \cdots x_{k-1}). \tag{17.34}$$

In other words, the null space of a symmetric k-linear operator K is the intersection of the null spaces of all of the linear operators $Kx_1 x_2 \cdots x_{k-1}$ which can be obtained from it by operation on $k - 1$ points $x_1, x_2, \ldots, x_{k-1} \in X$.

PROOF. If $x_0 \in N(K)$, then $Kx_0 = 0$, the zero $(k - 1)$-linear operator, so that

$$Kx_0 x_1 \cdots x_{k-1} = (Kx_1 x_2 \cdots x_{k-1})x_0 = 0 \tag{17.35}$$

for all $(x_1, x_2, \ldots, x_{k-1}) \in X^{k-1}$ and, thus,

$$N(K) \subset \bigcap_{(x_1, x_2, \ldots, x_{k-1}) \in X^{k-1}} N(Kx_1 x_2 \cdots x_{k-1}). \tag{17.36}$$

If now $x_0 \in \bigcap_{(x_1, x_2, \ldots, x_{k-1}) \in X^{k-1}} N(Kx_1 \dot{x}_2 \cdots x_{k-1})$, then

$$Kx_1 \cdots x_{k-1}x_0 = (Kx_0)x_1 \cdots x_{k-1} = 0 \tag{17.37}$$

for all $(x_1, x_2, \ldots, x_{k-1}) \in X^{k-1}$, so that Kx_0 is the zero $(k - 1)$-linear operator $x_0 \in N(K)$ and

$$\bigcap_{(x_1, x_2, \ldots, x_{k-1}) \in X^{k-1}} N(Kx_1 x_2 \cdots x_{k-1}) \subset N(K). \tag{17.38}$$

Equation 17.34 follows at once from (17.36) and (17.38).

The usefulness of this theorem is based on the fact that if

$$N(Kx_1 x_2 \cdots x_{k-1}) = \{0\}$$

for some choice of $x_1, x_2, \ldots, x_{k-1}$ we have $N(K) = \{0\}$. On the other hand, if $N(K)$ contains a nonzero vector x_0, then x_0 will be in the null space of any linear operator from X into Y of the form $L = Kx_1 x_2 \cdots x_{k-1}$.

Thus it follows that

$$B = \begin{pmatrix} 1 & -1 & -1 & 2 \\ 1 & 3 & 3 & 1 \end{pmatrix} \tag{17.39}$$

is a nonsingular bilinear operator in R^2, since for $x = (1, 1)$,

$$Bx = \begin{pmatrix} 1 & -1 & -1 & 2 \\ 1 & 3 & 3 & 1 \end{pmatrix}\begin{pmatrix} 1 \\ 1 \end{pmatrix} = \begin{pmatrix} 0 & 1 \\ 4 & 4 \end{pmatrix}, \tag{17.40}$$

is a nonsingular linear operator in R^2. However, B has no inverse, which is to be expected, since its range is a two-dimensional subspace of the four-dimensional space of 2×2 matrices. To see this note that

$$B\binom{\xi}{\eta} = \begin{pmatrix} \xi - \eta & -\xi + 2\eta \\ \xi + 3\eta & 3\xi + \eta \end{pmatrix} = \xi \begin{pmatrix} 1 & -1 \\ 1 & 3 \end{pmatrix} + \eta \begin{pmatrix} -1 & 2 \\ 3 & 1 \end{pmatrix}, \qquad (17.41)$$

so the range of B is the span of the linearly independent matrices

$$A_1 = \begin{pmatrix} 1 & -1 \\ 1 & 3 \end{pmatrix}, \qquad A_2 = \begin{pmatrix} -1 & 2 \\ 3 & 1 \end{pmatrix}, \qquad (17.42)$$

and is thus two-dimensional.

On the other hand, consider for $X = Y = C[0, 1]$ the following example of a bilinear operator represented by a kernel

$$B = B(s, t, u), \qquad (17.43)$$

continuous for $0 \leq s, t, u \leq 1$, where

$$Bx = \int_0^1 B(s, t, u)\, x(u)\, du, \qquad 0 \leq s, t \leq 1, \qquad (17.44)$$

is the kernel of a linear integral operator in $C[0, 1]$ for $x \in C[0, 1]$. For

$$B(s, t, u) = st^2 u^2, \qquad (17.45)$$

it follows that, if

$$x_0(s) = 4s - 3, \qquad (17.46)$$

then

$$Bx_0 = \int_0^1 st^2 u^2 (4u - 3)\, du = 0, \qquad 0 \leq s, t \leq 1, \qquad (17.47)$$

and thus $x_0 \in N(B)$. By Theorem 17.1, $x_0(s)$ will be an element of the nul space of any linear operator K in $C[0, 1]$ with a kernel which may be expressed in the form

$$Bx = K(s, t) = \int_0^1 st^2 u^2\, x(u)\, du, \qquad 0 \leq s, t \leq 1, \qquad (17.48)$$

for $x \in C[0, 1]$.

Multilinear operators may be used to define a simple class of nonlinear operators from X into Y; for example, if B is a bilinear operator from X into Y, then we may define the *quadratic operator* $Q(x)$ from X into Y by

$$Q(x) = Bxx \qquad (17.49)$$

for all $x \in X$. This operator is a natural generalization of the scalar function

$$q(x) = bx^2. \qquad (17.50)$$

There is no loss of generality in assuming that the bilinear operator B in (17.49) is symmetric, as

$$Bxx = B^*xx = \bar{B}xx \tag{17.51}$$

for all $x \in X$. Thus if B is nonsymmetric, then it may be replaced by its mean $\bar{B}$, which is symmetric, without altering the value of (17.49).

The introduction of the quadratic operator leads to the formulation of the abstract *quadratic equation*

$$Bxx + Lx + y = 0 \tag{17.52}$$

for $x \in X$, where we are given the symmetric bilinear operator B from X into Y, the linear operator L from X into Y, and the point $y \in Y$ [5]. Many important nonlinear equations of practical interest are of the form (17.52); for example, the partial differential equation $\Delta u = u^2$ [see (11.18)], the Riccati ordinary differential equation in Problem 4, Exercises 12 (see also Problem 3, Exercises 15), the integral equation (13.1) and its finite analog (13.40), and the system of reaction-rate equations in Problem 1, Exercises 16, are all quadratic equations of the form (17.52) between the appropriate Banach spaces. Additional examples of important quadratic equations will be introduced later. In spite of their simplicity, at least in comparison to other nonlinear problems, the theory of abstract quadratic equations is still in a primitive state, unlike the highly developed theory of linear equations.

If K is a k-linear operator from X into Y, we may define a nonlinear operator P from X into Y by

$$P(x) = Kx^k = K\overbrace{x \cdots x}^{k}, \tag{17.53}$$

which is obtained by setting $x_1 = x_2 = \cdots = x_k = x$. Again, K may be assumed to be symmetric. If A_k is a given k-linear operator from X into Y, $k = 1, 2, \ldots, n$, and $A_0 \in Y$ is given, the equation for $x \in X$,

$$A_n x^n + A_{n-1} x^{n-1} + \cdots + A_2 x^2 + A_1 x + A_0 = 0 \tag{17.54}$$

(nonlinear if $n \geq 2$), is called an *abstract polynomial equation of degree n in x*, where $A_2, A_3, \ldots, A_n$ are assumed to be symmetric. An adequate theory of this class for equations would advance the study of a number of important nonlinear problems.

EXERCISES 17

1. For $X = R^n$, $Y = R^m$, show that the bilinear operator B defined by (17.4) is symmetric if and only if

$$b_{ijk} = b_{ikj}, \qquad i = 1, 2, \ldots, m, \quad j, k = 1, 2, \ldots, n.$$

Give a similar condition for the k $-$ linear operator G given by (17.6) to be symmetric.

2. Derive an estimate, that is, an upper bound for $\|B\|$, where B is the bilinear integral operator in $C[0, 1]$ with the continuous kernel

$$B = B(s, t, u), \qquad 0 \leq s, t, u \leq 1.$$

3. Give necessary and sufficient conditions for the bilinear operator

$$B = (b_{111} \quad b_{112} \,|\, b_{121} \quad b_{122})$$

from R^2 into R to be singular. Do the same for the bilinear operator

$$B = \begin{pmatrix} b_{111} & b_{112} & b_{121} & b_{122} \\ b_{211} & b_{212} & b_{221} & b_{222} \end{pmatrix}$$

in R^2

4. If B is a bilinear operator from X into Y, show that the quadratic operator

$$Q(x) = Bxx$$

has the derivative

$$Q'(x_0) = Bx_0 + B^*x_0 = 2\bar{B}x_0.$$

5. Show that the *abstract polynomial operator*

$$P_n(x) = A_n x^n + \cdots + A_2 x^2 + A_1 x + A_0$$

appearing in (17.54) has the derivative

$$P_n'(x_0) = nA_n x_0^{n-1} + \cdots + 2A_2 x_0 + A_1,$$

assuming that $A_2, A_3, \ldots, A_n$ are symmetric.

6. Show than an operator B from X^2 into Y such that

$$B(\lambda_1 x_1 + \lambda_1' x_1', \lambda_2 x_2 + \lambda_2' x_2') = \lambda_1 \lambda_2 B(x_1, x_2) + \lambda_1 \lambda_2' B(x_1, x_2') + \\ + \lambda_1' \lambda_2 B(x_1', x_2) + \lambda_1' \lambda_2' B(x_1', x_2')$$

defines bilinear operators B, B^* from X into Y by

$$Bx = B(x, [\ ]), \quad B^*x = B([\ ], x),$$

provided a constant M exists such that $\|B(x, x')\| \leq M\|x\| \cdot \|x'\|$ for all $(x, x') \in X^2$. Is the operator B from X^2 into Y linear?

7. Investigate the solution of the quadratic equation

$$x = y + Bxx$$

in a Banach space X by the method of successive substitutions, starting from $x_0 = y$.

8. Show that any bilinear operator B from X into Y can be expressed as the sum of a symmetric bilinear operator, and a bilinear operator B_0 such that

$$B_0 xx = 0$$

for all $x \in X$.

18. SECOND AND HIGHER DERIVATIVES

Suppose that an operator P from X into Y is differentiable at x_0 and also at every point of the ball $U(x_0, r)$ for some $r > 0$. For each $x \in U(x_0, r)$, $P'(x_0)$ will be an element of the space $L(X, Y)$ of bounded linear operators from X into Y. Consequently, P' may be considered to be an operator which maps the ball $U(x_0, r)$ into some corresponding ball in the space $L(X, Y)$. By Definition 14.1, P' will be differentiable at x_0 if a bounded linear operator B from X into $L(X, Y)$ exists such that

$$\lim_{\|\Delta x\| \to 0} \frac{\|P'(x_0 + \Delta x) - P'(x_0) - B\,\Delta x\|}{\|\Delta x\|} = 0. \tag{18.1}$$

A bounded linear operator B from X into $L(X, Y)$ is, by Definition 17.2, a bilinear operator from X into Y. If the bilinear operator satisfying (18.1) exists, then it is called the *second derivative of P at x_0*, and

$$B = P''(x_0). \tag{18.2}$$

Thus the second derivative of an operator P from X into Y, obtained by differentiating the first derivative P' of P, is a bilinear operator from X into Y. Furthermore it is symmetric.

Theorem 18.1. If the second derivative $P''(x_0)$ of an operator P from X into Y exists at $x = x_0$, then it is a symmetric bilinear operator from X into Y.

PROOF.   The bilinearity of $P''(x_0)$ has already been established. To prove its symmetry, assume that $x, z \in X$ are any two nonzero vectors. Without loss of generality it may be assumed that $\|x\| = \|z\| = 1$. Suppose that $P'(x)$ exists in the ball $U(x_0, r)$ and that θ, θ' denote scalars with $|\theta| + |\theta'| < r$. By Definition 14.1

$$P'(x_0)z = \lim_{\theta \to 0} \frac{P(x_0 + \theta z) - P(x_0)}{\theta}, \tag{18.3}$$

which follows from (14.1) for the special case that $\Delta x = \theta z$. Similarly,

$$P'(x_0 + \theta' x)z = \lim_{\theta \to 0} \frac{P(x_0 + \theta' x + \theta z) - P(x_0 + \theta' x)}{\theta}. \tag{18.4}$$

Likewise, from (18.1), if $P''(x_0)$ exists, then

$$P''(x_0)xz = \lim_{\theta' \to 0} \frac{P'(x_0 + \theta' x)z - P'(x_0)z}{\theta'}. \tag{18.5}$$

Substitution of (18.3) and (18.4) into (18.5) yields

$$P''(x_0)xz = \lim_{\theta' \to 0} \lim_{\theta \to 0} \frac{P(x_0 + \theta' x + \theta z) - P(x_0 + \theta' x) - P(x_0 + \theta z) + P(x_0)}{\theta' \theta}. \tag{18.6}$$

Since by definition the derivative is independent of the manner in which

$\Delta x \to 0$, (18.6) may be rewritten

$$P''(x_0)xz = \lim_{\theta \to 0} \lim_{\theta' \to 0} \frac{P(x_0 + \theta z + \theta' x) - P(x_0 + \theta z) - P(x_0 + \theta' x) + P(x_0)}{\theta \theta'}$$

$$= \lim_{\theta \to 0} \frac{P'(x_0 + \theta z)x - P'(x_0)x}{\theta}$$

$$= P''(x_0)zx \qquad (18.7)$$

This proves the theorem.

For the case that $X = R_p{}^n$, $Y = R_p{}^m$ the second derivative has the form

$$P''(x_0) = \left(\frac{\partial^2 P_i}{\partial \xi_j \, \partial \xi_k} \right)_{x=x_0}, \qquad i = 1, 2, \dots, m, \quad j, k = 1, 2, \dots, n, \quad (18.8)$$

for $P(x) = (P_1(x), P_2(x), \dots, P_n(x))$, with the condition of symmetry implying that

$$\frac{\partial^2 P_i}{\partial \xi_j \, \partial \xi_k} = \frac{\partial^2 P_i}{\partial \xi_k \, \partial \xi_j}. \qquad (18.9)$$

For a reason to be explained later the convention

$$\frac{\partial^2 P_i}{\partial \xi_j \, \partial \xi_k} = \frac{\partial}{\partial \xi_k} \frac{\partial P_i}{\partial \xi_j} \qquad (18.10)$$

is adopted.

By extending these ideas it is possible to frame an inductive definition of higher derivatives of an operator P from a Banach space X into a Banach space Y.

Definition 18.1. If P is differentiable $k - 1$ times in some ball $U(x_0, r)$, $r > 0$, and a k-linear operator K from X into Y exists such that

$$\lim_{\|\Delta x\| \to 0} \frac{\| P^{(k-1)}(x_0 + \Delta x) - P^{(k-1)}(x_0) - K \Delta x \|}{\|\Delta x\|} = 0, \qquad (18.11)$$

then K is called the kth *derivative of P at x_0*, and

$$K = P^{(k)}(x_0). \qquad (18.12)$$

By essentially the same argument used in the proof of Theorem 18.1 it can be shown that if $P^{(k)}(x_0)$ exists it is symmetric.

If $L(x) = Lx$, with L a bounded linear operator from X into Y, then

$$L'(x_0) = L, \qquad L''(x_0) = 0, \qquad (18.13)$$

for all $x_0 \in X$.

More generally, consider the *abstract polynomial operator*

$$P_n(x) = A_n x^n + \cdots + A_2 x^2 + A_1 x + A_0 \tag{18.14}$$

from X into Y, where A_k, $k = 2, 3, \ldots, n$, is a symmetric k-linear operator from X into Y, A_1 is a bounded linear operator from X into Y, and $A_0 \in Y$. It is evident that

$$\begin{aligned}
P'_n(x_0) &= nA_n x_0^{n-1} + \cdots + 2A_2 x_0 + A_1, \\
P''_n(x_0) &= n(n-1)A_n x^{n-2} + \cdots + 6A_3 x_0 + 2A_2, \\
&\cdot \\
&\cdot \\
&\cdot \\
P_n^{(n)}(x_0) &= n!\, A_n, \\
P_n^{(n+1)}(x_0) &= P_n^{(n+2)}(x_0) = \cdots = 0.
\end{aligned} \tag{18.15}$$

As in elementary algebra [6], the value of $P_n(x)$ and its derivatives may be calculated at $x = x_0$ by *Horner's algorithm*, the computations being programmed according to Figure 18.1.

$$\begin{array}{llllll|l}
A_n & A_{n-1} & \cdots & A_1 & & A_0 & x \\
 & A_n x & \cdots & A_n x^{n-1} + \cdots + A_2 x & & A_n x^n + \cdots + A_1 x & \\
\hline
A_n & A_n x + A_{n-1} & \cdots & A_n x^{n-1} + \cdots + A_2 x + A_1 & \big| & A_n x^n + \cdots + A_1 x + A_0 = P_n(x) & \\
 & A_n x & \cdots & (n-1)A_n x^{n-1} + \cdots + A_2 x & \big| & & \\
\hline
A_n & 2A_n x + A_{n-1} & \cdots \big| & nA_n x^{n-1} + \cdots + 2A_2 x + A_1 = P_n{}'(x) & & & \\
 & A_n x & \cdots \big| & & & & \\
\hline
A_n & 3A_n x + A_{n-1} & \cdots & & & & \\
\cdots \big| & \cdots \quad \cdots & \cdots & & & & \\
\end{array}$$

$$A_n = \frac{1}{n!} P_n^{(n)}(x)$$

Figure 18.1. Horner's algorithm.

For any $z \in X$,

$$P_n(x_0 + z) = P_n(x_0) + P'_n(x_0)z + \tfrac{1}{2}P''_n(x_0)z^2 + \cdots + \frac{1}{n!}P_n^{(n)}(x_0)z^n, \tag{18.16}$$

a finite Taylor series. In particular,

$$A_k = \frac{1}{k!} P_n^{(k)}(0), \qquad k = 0, 1, \ldots, n, \tag{18.17}$$

with

$$P_n^{(0)}(0) = A_0, \qquad 0! = 1, \tag{18.18}$$

by definition.

For an operator P from X into Y such that

$$P(x_0), P'(x_0), P''(x_0), \ldots, P^{(n)}(x_0), \ldots \tag{18.19}$$

exist we may entertain the notion of an *abstract Taylor series*

$$P(x_0 + h) = \sum_{k=0}^{\infty} \frac{1}{k!} P^{(k)}(x_0) h^k \tag{18.20}$$

for $h \in X$, where we define

$$\frac{1}{0!} P^{(0)}(x_0) h^0 = P(x_0). \tag{18.21}$$

Suppose that

$$a_k \geq \|P^{(k)}(x_0)\|, \qquad k = 0, 1, 2, \ldots. \tag{18.22}$$

Then the scalar power series

$$p(\zeta) = \sum_{k=0}^{\infty} a_k \zeta^k \tag{18.23}$$

is called the *scalar majorant series* for the abstract Taylor series (18.20). It follows at once that, if (18.23) converges absolutely in the disk $|\zeta| < r$, then (18.20) converges for $h \in U_0(r)$. This is true because the partial sums

$$P_m(x_0 + h) = \sum_{k=0}^{m} \frac{1}{k!} P^{(k)}(x_0) h^k \tag{18.24}$$

of (18.20) form a Cauchy sequence, since

$$\|P_{m+p}(x_0 + h) - P_m(x_0 + h)\| \leq \sum_{k=m+1}^{m+p} a_k |\zeta|^k \tag{18.25}$$

and the absolute convergence of the series (18.23) guarantees the satisfaction of Definition 5.1 of a Cauchy sequence.

As usual, the supremum R of the values of r for which the scalar power series (18.23) converges in the disk $|\zeta| < r$ is called the *radius of convergence* of the power series. If (18.23) converges in every disk of finite radius, we take $R = \infty$. In general, the series (18.23) will be a scalar majorant series for the *abstract power series*

$$P(z) = \sum_{n=0}^{\infty} A_n z^n, \tag{18.26}$$

provided that

$$a_n \geq \|A_n\|, \tag{18.27}$$

where A_n, $n = 2, 3, \ldots$, are symmetric n-linear operators from X into Y and $A_0 z^0 = A_0$ is an element of Y.

Theorem 18.2. If the scalar majorant series (18.23) has radius of convergence R, then the abstract power series (18.26) converges in the ball $U_0(R)$ [if $R = \infty$, then $U_0(\infty) = X$]; furthermore, at any $z_0 \in U_0(R)$, P is differentiable an arbitrary number of times, with

$$P^{(m)}(z_0) = \sum_{n=m}^{\infty} n(n-1)\cdots(n-m+1)A_m z_0^{n-m}, \qquad (18.28)$$

where $A_m z^0 = A_m$.

PROOF. The convergence of (18.26) in $U_0(R)$ follows from the fact that the partial sums of (18.26) form a Cauchy sequence [see (18.25)]. Since a scalar power series is differentiable an arbitrary number of times in its disk of convergence [7], the series

$$p^{(m)}(\zeta_0) = \sum_{n=m}^{\infty} n(n-1)\cdots(n-m+1)a_n \zeta_0^{n-m} \qquad (18.29)$$

thus converges for $|\zeta| < R$ and is a scalar majorant series for (18.28), which consequently converges for $z_0 \in U_0(R)$. This completes the proof.

From (18.27) and (18.28)

$$A_k = \frac{1}{k!} P^{(k)}(0), \qquad (18.30)$$

By translation to x_0 the validity of the Taylor series expansion (18.24) of an operator P in the ball $U(x_0, R)$ is established, provided that (18.22) is satisfied, where R is the radius of convergence of the scalar majorant series (18.23). An operator P which has an abstract Taylor series expansion in a ball $U(x_0, r)$ is said to be *analytic* in $U(x_0, r)$.

The uniqueness of the power series expansion of an analytic operator in a ball $U(x_0, r)$ may be established in the same way as for a scalar series. This can be used to obtain higher derivatives of operators which have power series expansions. For example, in $L(X)$, if L_0^{-1} exists, then

$$P(L) = L^{-1} = \sum_{n=0}^{\infty} (-L_0^{-1}(L - L_0))^n L_0^{-1}, \qquad (18.31)$$

which converges for $L \in U(L_0, 1/\|L_0^{-1}\|)$ by Theorem 10.1. Setting $H = L - L_0$, we have

$$P'(L_0) = -L_0^{-1}[\quad]L_0^{-1}, \qquad (18.32)$$

as was found previously [see (15.32)]. For the representation of higher derivatives, it is convenient to introduce *indexed place holders* $[\quad]_1, [\quad]_2, \ldots$. For the operation of a formal operator on a string, the first (adjacent) variable occupies the place held by $[\quad]_1$, the second variable goes into the position indicated by [ ]$_2$, and so forth. Thus

$$\begin{pmatrix} A & [\ ]_2 & B \\ [\ ]_1 & C & D \end{pmatrix} XY = \begin{pmatrix} A & Y & B \\ X & C & D \end{pmatrix}. \tag{18.33}$$

With this convention it follows from (18.31) that

$$P''(L_0) = L_0^{-1}[\ ]_1 L_0^{-1}[\ ]_2 L_0^{-1} + L_0^{-1}[\ ]_2 L_0^{-1}[\ ]_1 L_0^{-1}, \tag{18.34}$$

a symmetric bilinear operator in $L(X)$. Of course, (18.34) could be obtained directly by differentiation of

$$P'(L) = -L^{-1}[\ ]L^{-1} \tag{18.35}$$

in the ball $U(L_0^{-1}, 1/\|L_0^{-1}\|)$. The placeholder [ ] in (18.35) is denoted by [ ]$_1$ in (18.34).

For $X = R_p{}^n$, $Y = R_p{}^m$, and $P(x) = (P_1(x), P_2(x), \ldots, P_m(x))$, we have

$$P^{(k)}(x_0) = \left(\frac{\partial^k P_i}{\partial \xi_{j_1} \partial \xi_{j_2} \cdots \partial \xi_{j_k}} \right)_{x=x_0},$$

$$i = 1, 2, \ldots, m, \quad j_1, j_2, \ldots, j_k = 1, 2, \ldots, n. \tag{18.36}$$

Even for fairly small values of m, n, k, the number of distinct components of $P^{(k)}(x_0)$ may be large enough to require special programming techniques in actual computational applications.

EXERCISES 18

1. In $L(X)$, the space of bounded linear operators in a Banach space X, we may define

$$\exp(L) = \sum_{n=0}^{\infty} \frac{L^n}{n!}$$

which converges in $L(X)$. Find

$$\exp'(0), \qquad \exp''(0).$$

Give a representation for

$$\exp'(L_0).$$

2. In $L(X)$ define

$$(I - L)^m$$

by an abstract power series. Find the radius of convergence of the resulting *abstract binomial series* and discuss its differentiability at $L = 0$ and $L = L_0$.
3. Find the first three derivatives of the operator

$$P\begin{pmatrix} \xi \\ \eta \end{pmatrix} = \begin{pmatrix} \xi^2 + \eta^2 - 4 \\ \xi \tan \eta \end{pmatrix}$$

in R^2.

4. Give a formula for the number of distinct components of $P^{(k)}(x_0)$ as given by (18.36), where $P(x)$ is an operator from R^n into R^m. Evaluate this number for $m = n = 2, 5, 10, 20, 100$.

5. Compute the second derivative of the Urysohn integral operator

$$U(x) = \int_0^1 f(s, t, x(t)) \, dt$$

in $C[0, 1]$.

6. Program Horner's algorithm for a digital computer, assuming that subroutines are available for calculating $A_k x$, where A_k is a k-linear operator from $X = R^n$ into $Y = R^m$, and $x \in R^n$.

19. HIGHER DERIVATIVES IN PRODUCT SPACES

To define higher partial derivatives of operators on a product space, the notion of a multilinear operator will be extended to situations involving more than two basic spaces. To illustrate the idea involved in this extension, suppose that P is an operator from $X \times Y$ into Z, that is,

$$z = P(x, y) \tag{19.1}$$

is an element of Z for each $x \in X, y \in Y$. On the basis of the concepts developed in Section 16, the partial derivative

$$P_x(x, y) = \frac{\partial P(x, y)}{\partial x} \tag{19.2}$$

will be a bounded linear operator from X into Z, that is, for each $x \in X$, $y \in Y$, the linear operator (19.2) is an element of $L(X, Z)$, the space of bounded linear operators from X into Z.

By the definition of the partial derivative of an operator, if $P_x(x, y)$ is differentiable with respect to x (for y fixed), then

$$P_{xx}(x, y) = \frac{\partial}{\partial x} P_x(x, y) = \frac{\partial^2 P(x, y)}{\partial x^2} \tag{19.3}$$

will be a bounded linear operator from X into $L(X, Z)$, that is, an element of $L(X, L(X, Z))$. Thus (19.3) is a bilinear operator from X into Z, as explained in Section 17. On the other hand, suppose that the operator (19.2) is differentiable with respect to y for x fixed. Then, by Definition 14.1, the *mixed second partial derivative*

$$P_{xy}(x, y) = \frac{\partial}{\partial y} \frac{\partial P(x, y)}{\partial x} = \frac{\partial^2 P(x, y)}{\partial x \, \partial y}, \tag{19.4}$$

if it exists, will be a linear operator from Y into $L(X, Z)$, that is, (19.4) will be an element of $L(Y, L(X, Z))$. Setting

$$B^* = P_{xy}(x^{(0)}, y^{(0)}) \tag{19.5}$$

for some fixed $x = x^{(0)}$, $y = y^{(0)}$, we have that B^*y is a bounded linear operator from X into Z for any $y \in Y$, and

$$B^*yx = (B^*y)x \tag{19.6}$$

is an element of Z for a given $x \in X$.

In the same way, for

$$P_y(x, y) = \frac{\partial P(x, y)}{\partial y}, \tag{19.7}$$

we may define

$$P_{yy}(x, y) = \frac{\partial}{\partial y} \frac{\partial P(x, y)}{\partial y} = \frac{\partial^2 P(x, y)}{\partial y^2}, \tag{19.8}$$

which is a linear operator from Y into $L(Y, Z)$, that is, a bilinear operator from Y into Z, and

$$P_{yx}(x, y) = \frac{\partial}{\partial x} \frac{\partial P(x, y)}{\partial y} = \frac{\partial^2 P(x, y)}{\partial y \, \partial x} \tag{19.9}$$

is a linear operator from X into $L(Y, Z)$. The notational convention concerning mixed partial derivatives adopted in (19.4) and (19.9) is intended to indicate that, for example,

$$\frac{\partial}{\partial y} \frac{\partial P(x, y)}{\partial x} = \frac{\partial^2 P(x, y)}{\partial x \, \partial y} \tag{19.10}$$

operates first on an element of Y then on an element of X to yield the element

$$z = \frac{\partial^2 P(x, y)}{\partial x \, \partial y} yx \tag{19.11}$$

of the space Z.

For

$$B = P_{yx}(x^{(0)}, y^{(0)}), \tag{19.12}$$

it follows by exactly the same reasoning as used in the proof of Theorem 18.1 that

$$Bxy = B^*yx \tag{19.13}$$

for all $x \in X, y \in Y$. Thus, although $B \in L(Y, L(X, Z))$ and $B^* \in L(X, L(Y, Z))$ are elements of different spaces if $X \neq Y$, the mixed second partial derivatives (19.5) and (19.12) have the symmetry property (19.13).

Combining the above results, we find that

$$P''(x^{(0)}, y^{(0)}) = \left(\frac{\partial^2 P}{\partial x^2} \quad \frac{\partial^2 P}{\partial x \, \partial y} \middle| \frac{\partial^2 P}{\partial y \, \partial x} \quad \frac{\partial^2 P}{\partial y^2} \right)_{(x,y)=(x^{(0)},y^{(0)})}, \tag{19.14}$$

a symmetric bilinear operator from $X \times Y$ into Z.

For $X = R^n$, $Y = R^m$, $Z = R^q$, and operator $B \in L(X, L(Y, Z))$ would have the representation

$$B = (b_{ijk}), \quad i = 1, 2, \ldots, q, \quad j = 1, 2, \ldots, m, \quad k = 1, 2, \ldots, n,$$

$$(19.15)$$

with $z = Bxy$ given by

$$\zeta_i = \sum_{j=1}^{m} \sum_{k=1}^{n} b_{ijk} \xi_k \eta_j, \quad i = 1, 2, \ldots, q, \quad (19.16)$$

where $z = (\zeta_1, \zeta_2, \ldots, \zeta_q), y = (\eta_1, \eta_2, \ldots, \eta_m)$, and $x = (\xi_1, \xi_2, \ldots, \xi_n)$. The operator

$$B^* = (b^*_{ijk}) \quad (19.17)$$

in $L(R^m, L(R^n, R^q))$ which satisfies (19.13) has the components

$$b^*_{ijk} = b_{ikj}, \quad i = 1, 2, \ldots, q, \quad j = 1, 2, \ldots, n, \quad k = 1, 2, \ldots, m.$$

$$(19.18)$$

If

$$P(x, y) = (P_1(x, y), P_2(x, y), \ldots, P_q(x, y)) \quad (19.19)$$

is an operator from $R^n \times R^m$ into $Z = R^q$, then we find by straightforward computation that

$$\frac{\partial^2 P}{\partial x^2} = \left(\frac{\partial^2 P_i}{\partial \xi_j \, \partial \xi_k} \right), \quad i = 1, 2, \ldots, q, \quad j, k = 1, 2, \ldots, n,$$

$$\frac{\partial^2 P}{\partial x \, \partial y} = \left(\frac{\partial^2 P_i}{\partial \xi_j \, \partial \eta_k} \right), \quad i = 1, 2, \ldots, q, \quad j = 1, 2, \ldots, n, \quad k = 1, 2, \ldots, m,$$

$$\frac{\partial^2 P}{\partial y \, \partial x} = \left(\frac{\partial^2 P_i}{\partial \eta_j \, \partial \xi_k} \right), \quad i = 1, 2, \ldots, q, \quad j = 1, 2, \ldots, m, \quad k = 1, 2, \ldots, n,$$

$$\frac{\partial^2 P}{\partial y^2} = \left(\frac{\partial^2 P_i}{\partial \eta_j \, \partial \eta_k} \right), \quad i = 1, 2, \ldots, q, \quad j, k = 1, 2, \ldots, m.$$

$$(19.20)$$

Higher partial derivatives in product spaces may be defined in much the same way as the foregoing examples for the second partial derivatives of an operator P from $X \times Y$ into Z. To develop a systematic notation, suppose that we are concerned with Banach spaces $X_1, X_2, \ldots$. Define

$$X_{ij} = L(X_j, X_i), \quad (19.21)$$

the space of bounded linear operators *from X_j into X_i*. Elements of X_{ij} are denoted by L_{ij}, etc. Similarly,

$$X_{ijk} = L(X_k, X_{ij}) = L(X_k, L(X_j, X_i)) \quad (19.22)$$

will denote the space of bounded *bilinear operators* B_{ijk} from X_k into X_{ij}. In general, let

$$X_{ij_1 j_2 \cdots j_k} = L(X_{j_k}, X_{ij_1 j_2 \cdots j_{k-1}}) \qquad (19.23)$$

denote the space of bounded linear operators from X_{j_k} into $X_{ij_1 j_2 \cdots j_{k-1}}$. The elements

$$K = K_{ij_1 j_2 \cdots j_k} \qquad (19.24)$$

of $X_{ij_1 j_2 \cdots j_k}$ are a natural generalization of the k-linear operators introduced in Section 17. There

$$X_i = Y, X_{j_1} = X_{j_2} = \cdots = X_{j_k} = X, \qquad (19.25)$$

a special case.

Suppose now that P_i is an operator from

$$X = \prod_{q=1}^{n} X_{j_q} \qquad (19.26)$$

into X_i, and that P_i has all partial derivatives of orders $1, 2, \ldots, k-1$ in some ball $U(x_0, r)$, where $r > 0$ and

$$x_0 = (x_{j_1}^{(0)}, x_{j_2}^{(0)}, \ldots, x_{j_n}^{(0)}) \qquad (19.27)$$

is a given element of X. To avoid proliferation of subscripts to the point that the notation becomes undecipherable, renumber the original spaces so that

$$j_1 = 1, j_2 = 2, \ldots, j_n = n. \qquad (19.28)$$

(These n spaces do not all have to be distinct.) Thus

$$x_0 = (x_1^{(0)}, x_2^{(0)}, \ldots, x_n^{(0)}). \qquad (19.29)$$

A typical partial derivative of order $k-1$ of P_i at x_0 is an operator

$$K_{i p_1 p_2 \cdots p_{k-1}} = \frac{\partial^{k-1} P_i(x_0)}{\partial x_{p_1} \partial x_{p_2} \cdots \partial x_{p_{k-1}}} \qquad (19.30)$$

in $X_{i p_1 p_2 \cdots p_{k-1}}$, where

$$1 \leq p_1, p_2, \ldots, p_{k-1} \leq n. \qquad (19.31)$$

Let $T(x_{p_k})$ denote the operator from X_{p_k} into $X_{i p_1 p_2 \cdots p_{k-1}}$ obtained from (19.30) by setting

$$x_j = x_j^{(0)}, \qquad j \neq p_k, \qquad (19.32)$$

for some p_k, $1 \leq p_k \leq n$. Then if

$$T'(x_{p_k}^{(0)}) = \frac{\partial}{\partial x_{p_k}} \frac{\partial^{k-1} P_i(x_0)}{\partial x_{p_1} \partial x_{p_2} \cdots \partial x_{p_{k-1}}} = \frac{\partial^k P_i(x_0)}{\partial x_{p_1} \partial x_{p_2} \cdots \partial x_{p_k}} \qquad (19.33)$$

exists, it will be an element of $X_{i p_1 p_2 \cdots p_k}$ and called the *partial derivative of order k of P_i with respect to* $x_{p_1}, x_{p_2}, \ldots, x_{p_k}$ *at* x_0.

If P_i is differentiable k times at x_0, that is, if all partial derivatives of order k of P_i exist at $x = x_0$, then

$$\frac{\partial^k P_i(x_0)}{\partial x_{p_1} \partial x_{p_2} \cdots \partial x_{p_k}} x_{p_1} x_{p_2} \cdots x_{p_k} = \frac{\partial^k P_i(x_0)}{\partial x_{q_1} \partial x_{q_2} \cdots \partial x_{q_k}} x_{q_1} x_{q_2} \cdots x_{q_k} \quad (19.34)$$

for any permutation $q_1, q_2, \ldots, q_k$ of the integers $p_1, p_2, \ldots, p_k$ and any choice of the points $x_{p_1}, x_{p_2}, \ldots, x_{p_k}$ from the respective spaces $X_{p_1}, X_{p_2}, \ldots, X_{p_k}$. This assertion may be proved in the same way as Theorem 18.1.

On the basis of these considerations, if $P = (P_1, P_2, \ldots, P_m)$ is an operator from $X = X_1 \times X_2 \times \cdots \times X_n$ into $Y = Y_1 \times Y_2 \times \cdots \times Y_m$, then the array

$$P^{(k)}(x_0) = \left(\frac{\partial^k P_i}{\partial x_{j_1} \partial x_{j_2} \cdots \partial x_{j_k}} \right)_{x=x_0},$$

$$i = 1, 2, \ldots, m, \quad j_1, j_2, \ldots, j_k = 1, 2, \ldots, n, \quad (19.35)$$

will represent the kth derivative of P at $x_0 = (x_1^{(0)}, x_2^{(0)}, \ldots, x_n^{(0)})$.

EXERCISES 19

1. Find the second derivative of the operator from $X \times \Lambda$ into itself defined by

$$y = Ax - \lambda x,$$
$$\eta = \phi x - 1,$$

where A is a bounded linear operator in X, $x \in X$, $\lambda \in \Lambda$, and ϕ is a bounded linear functional on X.

2. Suppose that a computer program is available that will compute the partial derivative of a given function with respect to a specified variable. Discuss the organization of a program which will calculate derivatives of an operator $P = (P_1, P_2, \ldots, P_m)$ from R^n into R^m. Also discuss calculation of the Taylor series

$$P(x_0 + h) = \sum_{n=0}^{\infty} \frac{1}{n!} P^{(n)}(x_0) h^n$$

to a specified number of terms. The organization of these programs may be shown in the form of a flow chart or in terms of statements in some computer programming language.

3. Calculate the first four derivatives of the operator P in R^2 defined by

$$P\begin{pmatrix} \xi \\ \eta \end{pmatrix} = \begin{pmatrix} \xi^2 + \sin \xi\eta + \eta^2 \\ 1 + \dfrac{1}{1 + \xi^3} - 4\eta^5\xi^3 \end{pmatrix}.$$

4. If X_1, X_2, X_3 are Banach spaces over the scalar field Λ, then an operator B from $X_1 \times X_2$ into X_3 is said to be *linear with respect to each variable* if

$$B(ax_1 + by_1, cx_2 + dy_2) = acB(x_1, x_2) + adB(x_1, y_2) + bcB(y_1, x_2) + bdB(y_1, y_2)$$

for all $x_1, y_1 \in X_1$, $x_2, y_2 \in X_2$, and scalars a, b, c, d, and *bounded jointly with respect to both variables* if a positive real number b exists such that

$$\|B(x_1, x_2)\| \le b\|x_1\| \cdot \|x_2\|.$$

For $x_1 = x_1^{(0)}$ fixed define the operator B_{32} by

$$B_{32}x_2 = B(x_1^{(0)}, x_2)$$

and the operator B_{321} by

$$B_{321}x_1^{(0)} = B_{32}.$$

Similarly, define B_{31} and B_{312}. Show that $B_{32} \in X_{32}$, $B_{321} \in X_{321}$, $B_{31} \in X_{31}$, $B_{321} \in X_{321}$, and that

$$B_{321}x_1x_2 = B_{312}x_2x_1 = B(x_1, x_2)$$

for all $x_1 \in X_1$, $x_2 \in X_2$.

20. RIEMANN INTEGRATION AND TAYLOR'S THEOREM

This section is based upon the results in the fundamental paper by L. M. Graves [8]. The theory presented here will be limited to a form suitable for later applications, rather than being stated in the greatest possible generality.

One difficulty that faces the functional analyst, hence the numerical analyst dealing with computational solution (or approximate solution) of problems posed in Banach spaces, is that the *mean value theorem* for differentiable real functions $f = f(x)$,

$$f(b) - f(a) = f'(\theta)(b - a), \tag{20.1}$$

where θ is some number satisfying $a < \theta < b$, does not hold in the more general setting of operators between Banach spaces. To see this let

$$P\begin{pmatrix} \xi \\ \eta \end{pmatrix} = \begin{pmatrix} \xi(\eta - 1) \\ \eta^2(\xi - 1) \end{pmatrix} \tag{20.2}$$

in R^2. Obviously

$$P\begin{pmatrix} 0 \\ 0 \end{pmatrix} = P\begin{pmatrix} 1 \\ 1 \end{pmatrix} = 0. \tag{20.3}$$

The derivative of the operator P defined by (20.2) is

$$P'\begin{pmatrix} \xi \\ \eta \end{pmatrix} = \begin{pmatrix} \eta - 1 & \xi \\ \eta^2 & 2\eta(\xi - 1) \end{pmatrix}. \tag{20.4}$$

For $a = (0, 0)$, $b = (1, 1)$,

$$P'\begin{pmatrix} \xi \\ \eta \end{pmatrix}(b - a) = \begin{pmatrix} \xi + \eta - 1 \\ \eta^2 + 2\eta(\xi - 1) \end{pmatrix}. \tag{20.5}$$

The line from $(0, 0)$ to $(1, 1)$ has the representation

$$\xi = \theta, \qquad \eta = \theta, \qquad 0 \leq \theta \leq 1. \tag{20.6}$$

From (20.5), since

$$\begin{pmatrix} 2\theta - 1 \\ 3\theta^2 - 2\theta \end{pmatrix} \neq 0, \qquad 0 < \theta < 1, \tag{20.7}$$

there is no direct analog to (20.1), even in a space as simple as R^2.

On the other hand, (20.1) implies that

$$|f(b) - f(a)| \leq \sup_{0 \leq \theta \leq 1} |f'(\theta)| \cdot |b - a|, \tag{20.8}$$

a possibly useful estimate for the magnitude of the difference $f(b) - f(a)$. It turns out that a generalization of (20.8) does hold for a differentiable operator P from a Banach space X into a Banach space Y, namely,

$$\|P(x_1) - P(x_0)\| \leq \sup_{x \in L(x_0, x_1)} \|P'(\bar{x})\| \cdot \|x_1 - x_0\|, \tag{20.9}$$

where

$$L(x_0, x_1) = \{x : x = \theta x_1 + (1 - \theta)x_0, 0 \leq \theta \leq 1\} \tag{20.10}$$

is the *line segment joining the points* x_0 *and* x_1 (see Problem 1, Exercises 5).

To prove (20.9) it is convenient to extend the concept of the Riemann integral to abstract functions of a real variable, that is, operators which map R into a Banach space Y. If P is an operator from a Banach space X into a Banach space Y, then for

$$x(\theta) = \theta x_1 + (1 - \theta)x_0, \qquad 0 \leq \theta \leq 1, \tag{20.11}$$

the abstract function

$$P(\theta) = P(x(\theta)) = P(\theta x_1 + (1 - \theta)x_0) \tag{20.12}$$

will map the real interval $0 \leq \theta \leq 1$ into an *abstract arc* in Y, which starts at $y_0 = P(x_0)$ and ends at $y_1 = P(x_1)$.

As in the usual definition of the Riemann integral, we partition the interval $0 \leq \theta \leq 1$ into n subintervals of lengths $\Delta\theta_i, i = 1, 2, \ldots, n$, choose points θ_i interior to the corresponding subinterval, and form the sum

$$\sum_{\pi} P(\theta_i)\,\Delta\theta_i = \sum_{i=1}^{n} P(\theta_i)\,\Delta\theta_i, \tag{20.13}$$

where π denotes the given partition of the interval. For any partition π, let

$$|\pi| = \max_{(i)} \Delta\theta_i. \tag{20.14}$$

Definition 20.1.   If

$$J = \lim_{|\pi| \to 0} \sum_{\pi} P(\theta_i)\,\Delta\theta_i \tag{20.15}$$

exists, then it is called the *Riemann integral of* $P(\theta)$ *from* 0 *to* 1, and is denoted by

$$J = \int_0^1 P(\theta)\, d\theta. \tag{20.16}$$

For the fixed path of integration along the line segment (20.11) it will sometimes be convenient to write

$$\int_{x_0}^{x_1} P(x)\, dx = \int_0^1 P(\theta)\, d\theta, \tag{20.17}$$

and call (20.17) the *integral of* $P(x)$ *from* x_0 *to* x_1.

The abstract Riemann integral defined in the foregoing has many properties in common with the ordinary scalar Riemann integral [8]; for example,

$$\int_0^1 (P(\theta) + Q(\theta))\, d\theta = \int_0^1 P(\theta)\, d\theta + \int_0^1 Q(\theta)\, d\theta \tag{20.18}$$

for integrable abstract functions $P(\theta)$, $Q(\theta)$, and, if $p(\theta)$ is an integrable *scalar majorant function* for $P(\theta)$, that is, if

$$\|P(\theta)\| \le p(\theta), \qquad 0 \le \theta \le 1. \tag{20.19}$$

then

$$\left\| \int_0^1 P(\theta)\, d\theta \right\| \le \int_0^1 p(\theta)\, d\theta. \tag{20.20}$$

The integral on the right-hand side of (20.20) is, of course, an ordinary Riemann integral.

A sufficient condition for Riemann integrability of a bounded abstract function $P(\theta)$ can be shown to be that the set D_P of points of discontinuity of $P(\theta)$ forms a subset of *measure zero* of the interval $0 \le \theta \le 1$ [8]. This means the following: given any $\epsilon > 0$, there exists a countable collection of sub-intervals of $[0, 1]$ with lengths $\epsilon_1, \epsilon_2, \ldots$, the union of which contains every point of D_P and

$$\sum_{i=1}^{\infty} \epsilon_i < \epsilon. \tag{20.21}$$

Definition 20.2. A bounded abstract function $P(\theta)$ on $[0, 1]$ such that its set D_P of points of discontinuity is of measure zero is said to be *integrable* on $[0, 1]$.

If the operator P is differentiable at each $x \in L(x_0, x_1)$, then it follows from (20.11) and the chain rule (14.8) that

$$P'(\theta) = P'(\theta x_1 + (1 - \theta)x_0)(x_1 - x_0) \tag{20.22}$$

is an element of Y, so that $P'(\theta)$ is also an abstract function from $[0, 1]$ into Y. The relationship between an abstract function and the integral of its derivative is given by the following theorem.

Theorem 20.1. If $P'(\theta)$ is integrable, then

$$P(1) - P(0) = \int_0^1 P'(\theta)\, d\theta. \tag{20.23}$$

PROOF.   For simplicity set

$$\theta_i = (i - 1)\frac{1}{n}, \qquad \Delta\theta_i = \frac{1}{n}, \qquad i = 1, 2, \ldots, n, \tag{20.24}$$

for n a positive integer. Then

$$P(1) - P(0) - \sum_{i=1}^n P'(\theta_i)\Delta\theta_i = \sum_{i=1}^n \{P(\theta_i + \Delta\theta_i) - P(\theta_i) - P'(\theta_i)\Delta\theta_i\}. \tag{20.25}$$

By the definition of the derivative, given any $\epsilon > 0$, an n_0 exists such that

$$\|P(\theta_i + \Delta\theta_i^\cdot) - P(\theta_i) - P'(\theta_i)\Delta\theta_i\| < \frac{\epsilon}{2n} \tag{20.26}$$

for $n > n_0$, by using (20.24). Hence from (20.25),

$$\left\| P(1) - P(0) - \sum_{i=1}^n P'(\theta_i)\Delta\theta_i \right\| < \frac{\epsilon}{2} \tag{20.27}$$

for $n > n_0$. Also by (20.15), an n_1 exists such that

$$\left\| \sum_{i=1}^n P'(\theta_i)\Delta\theta_i - \int_0^1 P'(\theta)\, d\theta \right\| < \frac{\epsilon}{2} \tag{20.28}$$

for $n > n_1$. The combination of (20.27) and (20.28) for $n > \max\{n_0, n_1\}$ gives

$$\left\| P(1) - P(0) - \int_0^1 P'(\theta)\, d\theta \right\| < \epsilon, \tag{20.29}$$

which proves (20.23), since ϵ is arbitrarily small.

This proof makes use of the fact that the integrability of $P'(\theta)$ implies that the limit (20.15) exists for all suitable partitions of $[0, 1]$, which means that the limit for the particular partition defined by (20.24) exists and is equal to the integral of $P'(\theta)$.

As an example, consider $P(x)$ for $x = (\xi, \eta)$ in R^2 defined by (20.2). For $x_0 = (0, 0)$, $x_1 = (1, 1)$

$$P'(\theta) = \begin{pmatrix} \theta - 1 & \theta \\ \theta^2 & 2\theta(\theta - 1) \end{pmatrix}\begin{pmatrix} 1 \\ 1 \end{pmatrix} = \begin{pmatrix} 2\theta - 1 \\ 3\theta^2 - 2\theta \end{pmatrix}, \tag{20.30}$$

from (20.5) and (20.6). It follows that

$$P(1) - P(0) = \int_0^1 P'(\theta)\, d\theta = \int_0^1 \left(\frac{2\theta - 1}{3\theta^2 - 2\theta} \right) d\theta = \left. \left(\frac{\theta^2 - \theta}{\theta^3 - \theta^2} \right) \right|_0^1 = 0, \quad (20.31)$$

as required by Theorem 20.1.

Corollary 20.1. If $P'(x)$ is integrable from x_0 to x_1, then inequality (20.9) holds.

PROOF. By Theorem 20.1, inequality (20.20), and (20.22),

$$\|P(x_1) - P(x_0)\| = \left\| \int_0^1 P'(\theta)\, d\theta \right\| = \left\| \int_0^1 P'(\theta x_1 + (1 - \theta)x_0)(x_1 - x_0)\, d\theta \right\|$$

$$\leq \sup_{\bar{x} \in L(x_0, x_1)} \|P'(\bar{x})\| \cdot \|x_1 - x_0\|, \quad (20.32)$$

which is equivalent to (20.9).

Corollary 20.2 (*Integration by Parts*). If $P(\theta)$ is an abstract function and $Q(\theta)$ is a scalar function such that $P'(\theta)$ and $Q'(\theta)$ are integrable, then

$$P(1)\, Q(1) - P(0)\, Q(0) = \int_0^1 P(\theta)\, Q'(\theta)\, d\theta + \int_0^1 P'(\theta)\, Q(\theta)\, d\theta. \quad (20.33)$$

PROOF. Equation (20.33) follows at once from Theorem 20.1 for

$$F(\theta) = P(\theta)\, Q(\theta), \quad (20.34)$$

since

$$F'(\theta) = P(\theta)\, Q'(\theta) + P'(\theta)\, Q(\theta) \quad (20.35)$$

for a multiplicative product [see the derivation of formula (17.23)].

By repeated applications of the chain rule, the higher derivatives of $P(\theta)$ as defined by (20.12) are found to be

$$P^{(k)}(\theta) = P^{(k)}(\theta x_1 + (1 - \theta)x_0)(x_1 - x_0)^k, \quad k = 1, 2, \ldots. \quad (20.36)$$

Each of these derivatives will be an abstract function of θ.

Theorem 20.2 (*Taylor's Theorem*). Suppose that $P(x)$ is differentiable n time in the ball $U(x_0, r)$, $r > 0$, and $P^{(n)}(x)$ is integrable from x_0 to any $x_1 \in U(x_0, r)$; then

$$P(x_1) = P(x_0) + \sum_{k=1}^{n-1} \frac{1}{k!} P^{(k)}(x_0)(x_1 - x_0)^k + R_n(x_0, x_1), \quad (20.37)$$

where

$$R_n(x_0, x_1) = \int_0^1 P^{(n)}(\theta x_1 + (1 - \theta)x_0)(x_1 - x_0)^n \frac{(1 - \theta)^{n-1}}{(n - 1)!}\, d\theta. \quad (20.38)$$

PROOF. This theorem is true for $n = 1$ by Theorem 20.1, with (20.37) and (20.38) following immediately from (20.23). Assume that $P(x)$ is differentiable at least m times in $U(x_0, r)$, $m \geq 2$, and that (20.37) and (20.38) hold for $n = m - 1$. For the scalar function

$$Q(\theta) = \frac{(1 - \theta)^{m-1}}{(m - 1)!}, \tag{20.39}$$

we have

$$Q'(\theta) = -\frac{(1 - \theta)^{m-2}}{(m - 2)!}. \tag{20.40}$$

Also, if

$$F(\theta) = P^{(m-1)}(\theta x_1 + (1 - \theta)x_0)(x_1 - x_0)^{m-1}, \tag{20.41}$$

then

$$F'(\theta) = P^{(m)}(\theta x_1 + (1 - \theta)x_0)(x_1 - x_0)^{m}. \tag{20.42}$$

Hence from (20.38), (20.40), and (20.41),

$$R_{m-1}(x_0, x_1) = -\int_0^1 F(\theta)\, Q'(\theta)\, d\theta. \tag{20.43}$$

Substitution of (20.43) into formula (20.33) for integration by parts yields

$$R_{m-1}(x_0, x_1) = -F(1)\, Q(1) + F(0)\, Q(0) + \int_0^1 F'(\theta)\, Q(\theta)\, d\theta$$

$$= \frac{1}{(m - 1)!}\, P^{(m-1)}(x_0)(x_1 - x_0)^{m}$$

$$+ \int_0^1 P^{(m)}(\theta x_1 + (1 - \theta)x_0)(x_1 - x_0)^{m} \frac{(1 - \theta)^{m-1}}{(m - 1)!}\, d\theta$$

$$= \frac{1}{(m - 1)!}\, P^{(m-1)}(x_0)(x_1 - x_0)^{m} + R_m(x_0, x_1). \tag{20.44}$$

This establishes (20.37) by mathematical induction, which proves the theorem.

Corollary 20.3. If $P(x)$ satisfies the hypotheses of Theorem 20.2 in $U(x_0, r)$, then

$$\left\| P(x_1) - \sum_{k=0}^{n-1} \frac{1}{k!}\, P^{(k)}(x_0)(x_1 - x_0)^{k} \right\| \leq \sup_{\bar{x} \in L(x_0, x_1)} \| P^{(n)}(\bar{x}) \| \frac{\| x_1 - x_0 \|^{n}}{n!}. \tag{20.45}$$

PROOF. Inequality (20.45) follows immediately from (20.37), (20.38), and (20.20) with

$$p(\theta) = \frac{(1 - \theta)^{n-1}}{(n - 1)!} \sup_{\bar{x} \in L(x_0, x_1)} \| P^{(n)}(\bar{x}) \| \cdot \| x_1 - x_0 \|^{n}. \tag{20.46}$$

It follows from (20.45) that $P(x)$ will be analytic in the open ball $U(x_0, r)$, in the sense defined in Section 18, if

$$\lim_{n \to \infty} \| P^{(n)}(x) \| = 0 \tag{20.47}$$

uniformly in $U(x_0, r)$, assuming that $P(x)$ has derivatives of all orders.

Ordinarily the estimates obtained by using (20.9) or (20.45) will be somewhat crude, but may be adequate for a given application. With more computational effort, more precise estimates of $\| R_n(x_0, x_1) \|$ can be obtained, if necessary, by finding a function $q(\theta)$ such that

$$q(\theta) \geq \| P^{(m)}(\theta x_1 - (1 - \theta)x_0)(x_1 - x_0)^n \| \frac{(1 - \theta)^{n-1}}{(n - 1)!} \tag{20.48}$$

with less error than (20.46), and then evaluating

$$\int_0^1 q(\theta) \, d\theta \geq \| R_n(x_0, x_1) \|. \tag{20.49}$$

To illustrate this consider the function $P(x)$ in R_∞^2 defined by (20.2). $P'(x)$ is given by (20.4), and

$$P''(x) = \begin{pmatrix} 0 & 1 & \bigg| & 1 & 0 \\ 0 & 2\eta & \bigg| & 2\eta & 2(\xi - 1) \end{pmatrix}. \tag{20.50}$$

For $x_0 = (0, 0)$, $x_1 = (1, 1)$,

$$\sup_{\bar{x} \in L(x_0, x_1)} \| P''(\bar{x}) \| \leq 4 \tag{20.51}$$

(see Problem 5, Exercises 8), which gives

$$\| R_2(x_0, x_1) \| \leq 2 \tag{20.52}$$

by (20.45). On the other hand,

$$P'' \begin{pmatrix} \theta \\ \theta \end{pmatrix} \begin{pmatrix} 1 \\ 1 \end{pmatrix} \begin{pmatrix} 1 \\ 1 \end{pmatrix} = \begin{pmatrix} 2 \\ 6\theta - 2 \end{pmatrix}. \tag{20.53}$$

From (20.52),

$$\left\| P'' \begin{pmatrix} \theta \\ \theta \end{pmatrix} \begin{pmatrix} 1 \\ 1 \end{pmatrix} \begin{pmatrix} 1 \\ 1 \end{pmatrix} \right\| = \begin{cases} 2, & 0 \leq \theta \leq \frac{2}{3}, \\ 6\theta - 2, & \frac{2}{3} \leq \theta \leq 1, \end{cases} \tag{20.54}$$

which gives

$$\| R_2(x_0, x_1) \| \leq \int_0^{2/3} 2(1 - \theta) \, d\theta + \int_{2/3}^1 2(4\theta - 1 - 3\theta^2) \, d\theta = \frac{28}{27} \tag{20.55}$$

by (20.49)

EXERCISES 20

1. In $L(X)$ integrate the operator

$$P(L) = L^{-1}$$

from $L_0 = I$ to $L_1 = T$, where $\|I - T\| < 1$. Call the resulting function $\ln(T)$. Calculate $\exp[\ln(T)]$ (the function $\exp(L)$ was defined in Problem 1, Exercises 18).

2. Give an estimate for the maximum error involved in the approximation of the operator

$$P(x) = \begin{pmatrix} \xi + \sin \xi\eta + 1 \\ \xi^2 + \eta^2 - 4\xi\eta \end{pmatrix}$$

in κ_∞^2 by the first three terms of its Taylor series expansion in the ball $U_0(1) = U(0, 1)$, i.e., for $x_0 = 0$.

3. Show that, if P is an operator in a Banach space X such that $\|P''(x)\| \leq K$, $P''(x)$ is integrable, and $[P'(x_0)]^{-1}$ exists, then $[P'(x)]^{-1}$ exists in the ball $U(x_0, 1/K \|[P'(x_0)]^{-1}\|)$.

REFERENCES AND NOTES

[1] Maurice Fréchet. La notion de différentielle dans l'analyse générale. *Ann. Sci. École Norm. Sup.*, 3 (1925), 293–323.

[2] W. A. Granville, P. F. Smith, and W. R. Longley. *Elements of the Differential and Integral Calculus.* Rev. ed. Ginn, Boston, 1941.

[3] E. K. Blum. A Formal System for Differentiation. *J. Assoc. Comput. Mach.*, 13 (1966), 495–504.

[4] J. B. Rosser. *Logic for Mathematicians.* McGraw-Hill, New York, 1953. The concept of function is discussed on pp. 305–329.

[5] L. B. Rall. Quadratic Equations in Banach Spaces. *Rend. Circ. Mat. Palermo*, 10 (1961), 314–332.

[6] C. C. MacDuffee. *Theory of Equations.* Wiley, New York, 1954.

[7] Konrad Knopp. *Infinite Sequences and Series.* Tr. by Frederick Bagemihl. Dover, New York, 1956.

[8] L. M. Graves. Riemann Integration and Taylor's Theorem in General Analysis. *Trans. Amer. Math. Soc.*, 29 (1927), 163–177.

4

NEWTON'S METHOD AND
ITS APPLICATIONS

Newton's method for solving scalar algebraic and transcendental equations [1] is well known for its simplicity and effectiveness. As generalized to systems of equations by Ostrowski [2], and to operator equations in Banach spaces by L. V. Kantorovič [3], this method provides a powerful tool for the theoretical as well as the numerical investigation of nonlinear operator equations. In this chapter the theory of Newton's method as developed by L. V. Kantorovič will be presented, together with some of its applications to important types of nonlinear equations.

21. LINEARIZATION AND ITERATION

The problem considered here is to find a solution $x = x^*$ of the equation

$$P(x) = 0, \tag{21.1}$$

where P is an operator in a Banach space X, or at least an approximation to x^* which is sufficiently accurate for some given purpose. One way to approach this problem would be to find a linear equation

$$Lx = y \tag{21.2}$$

which approximates (21.1) in some sense, at least in a neighborhood of x^*. A solution of (21.2), found by the methods presented in Chapter 1 or elsewhere [2,3,4], would give an approximation to a solution of the original nonlinear equation (21.1). This approximate solution is said to be found by *linearization* of (21.1).

It is easy to construct a linear equation (21.2) which approximates the nonlinear equation (21.1) if the operator P is differentiable. By (14.10),

$$P(x) = P(x_0) + P'(x_0)(x - x_0) + \eta(x, x_0), \tag{21.3}$$

where

$$\|\eta(x, x_0)\| = o(\|x - x_0\|) \tag{21.4}$$

as $\|x - x_0\| \to 0$. The notation $o(\lambda)$ as $\lambda \to 0$ refers to any scalar function $f(\lambda)$ such that

$$\lim_{\lambda \to 0} \frac{f(\lambda)}{\lambda} = 0. \tag{21.5}$$

If $P(x^*) = 0$, that is, if x^* is a solution of (20.1), then it follows from (21.3) that

$$P(x_0) + P'(x_0)(x^* - x_0) = -\eta(x^*, x_0). \tag{21.6}$$

If x^* is close to x_0, then by neglecting the presumably extremely small quantity $\eta(x^*, x_0)$ we obtain the linear equation

$$P(x_0) + P'(x_0)(x - x_0) = 0. \tag{21.7}$$

The solution $x = x_1$ of this equation, if it exists and is unique, could be expected to be a fairly good approximation to x^*, with

$$P(x_1) = \eta(x_1, x_0) \tag{21.8}$$

from (21.3). Equation (21.7) may be written in the form (21.2), with

$$L = P'(x_0), \qquad y = P'(x_0)x_0 - P(x_0). \tag{21.9}$$

Equation (21.7) is said to be obtained from (21.1) by the process of *linearization by differentiation*, sometimes called *quasilinearization* or the *tangent method*, since (21.7) is the equation of the tangent line to the curve $y = P(x)$ at the point $(x_0, P(x_0))$ in the case that $P(x)$ is a real scalar function. (See Figure 21.1.) Equation (21.7) will have a unique solution x_1 if $[P'(x_0)]^{-1}$ exists, and we may write

$$x_1 = x_0 - [P'(x_0)]^{-1} P(x_0). \tag{21.10}$$

The advantage to linearization of a nonlinear equation is that the highly developed theory and machinery for solving linear equations or inverting linear operators can be brought to bear on (21.2). Of course, after (21.2) is solved, an estimate of the error of its solution as an approximation to a solution x^* of (21.1) must be provided if a linearization technique is to be of practical importance.

Equation (21.10) expresses x_1 as the sum of x_0 and the *Newton correction*

$$\Delta x_0 = -[P'(x_0)]^{-1} P(x_0). \tag{21.11}$$

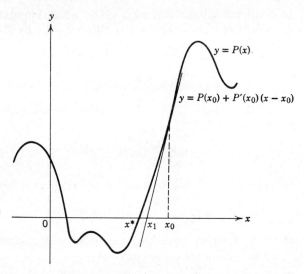

Figure 21.1   Linearization by differentiation in the real scalar case.

The Newton correction can also be obtained by solving the linear equation

$$P'(x_0) \, \Delta x_0 = -P(x_0), \tag{21.12}$$

which can possibly be done more simply than the inversion of the linear operator $P'(x_0)$.

The concept of *iteration* proceeds from the idea that, if x_0 is close to x^*, then x_1 is possibly closer, and the foregoing procedure may be repeated with x_0 replaced by x_1 to obtain a successive approximation x_2 to x^*, and so on. If $[P'(x_m)]^{-1}$ exists for $m = 0, 1, 2, \ldots$, we may define the *Newton sequence* $\{x_m\}$ by

$$x_{m+1} = x_m - [P'(x_m)]^{-1} P(x_m), \tag{21.13}$$

which corresponds to (21.11), or equivalently, by

$$P'(x_m)x_{m+1} = P'(x_m)x_m - P(x_m), \tag{21.14}$$

or

$$x_{m+1} = x_m + \Delta x_m, \qquad P'(x_m) \, \Delta x_m = -P(x_m), \qquad m = 0, 1, 2, \ldots. \tag{21.15}$$

The iterative process for construction of the Newton sequence $\{x_m\}$ by the use of (21.13), (21.14), or (21.15) is called *Newton's method* for solving the equation $P(x) = 0$.

In practice one of the three formulations of Newton's method previously given may have a computational advantage over the others. For theoretical purposes it will be assumed that the Newton sequence is generated by (21.13).

In common with the method of successive substitutions considered in Chapter 2, Newton's method has the computational feature that the same mathematical operations which produce x_1 from x_0 will give x_2 by substituting x_1 for x_0, and so forth, assuming that the formula used does not break down. Consequently, Newton's method is ideally suited for automatic execution by an electronic digital computer, at least for operators of arithmetic type and operators which can be approximated accurately by arithmetic operators.

It should be mentioned that the class of operators P to which Newton's method applies is restricted by the assumption that P is differentiable, which was not required for the method of successive substitutions. The class of differentiable operators is large enough, however, to include many examples of practical importance, as shown later.

Before discussing the theory of the convergence of Newton's method, several examples of scalar equations will be given to illustrate the linearization of the nonlinear equation (21.1) and the formation of the Newton sequence $\{x_m\}$.

For the scalar equation

$$f(x) = 0, \tag{21.16}$$

"inversion of the linear operator $f'(x_m)$" means simply division by the number $f'(x_m)$. Thus the Newton sequence is defined by

$$x_{m+1} = x_m - \frac{f(x_m)}{f'(x_m)}, \qquad m = 0, 1, 2, \ldots, \tag{21.17}$$

unless $f'(x_k) = 0$ for some value of k. As one particular example, consider

$$f(x) = x^2 - 3. \tag{21.18}$$

In this case (21.17) becomes

$$x_{m+1} = x_m - \frac{x_m{}^2 - 3}{2x_m} = \frac{1}{2}\left(x_m + \frac{3}{x_m}\right), \tag{21.19}$$

which is exactly Heron's method for calculating $\sqrt{3}$, discussed previously in Sections 11 and 12. More generally, consider the calculation of a kth root x of a number N. This can be done by solving an equation of the form $f(x) = 0$, with

$$f(x) = x^k - N. \tag{21.20}$$

From (21.20),

$$f'(x) = kx^{k-1}, \tag{21.21}$$

and (21.17) becomes

$$x_{m+1} = \frac{k-1}{k}\, x_m + \frac{N}{kx_m^{k-1}}. \tag{21.22}$$

Suppose that we wish to calculate the complex cube root of 3 in the second quadrant of the complex plane. By a simple graphical construction

$$x_0 = -1 + i \qquad (21.23)$$

appears to be fairly close to the value desired. The values of

$$x_m = \xi_m + i\eta_m \qquad (21.24)$$

are given in Table 21.1 for $n = 0, 1, 2, 3, 4$. The value of x_5 agreed with x_4 to seven decimal places [5]. Hence

$$x_4 = -0.7211248 + 1.2490248i \qquad (21.25)$$

is the most accurate approximation to the desired cube root that can be obtained by carrying eight significant digits in the calculation.

TABLE 21.1

Solution of $x^3 = (\xi + i\eta)^3 = 3$ by

Newton's method

m	ξ_m	η_m
0	−1.0000000	1.0000000
1	−0.6666667	1.1666667
2	−0.7256279	1.2549375
3	−0.7211515	1.2490519
4	−0.7211248	1.2490248

EXERCISES 21

1. Use (21.22) to calculate the fifth root of −1 lying in the first quadrant of the complex plane to five decimal place accuracy.
2. For the abstract quadratic equation

$$Q(x) \equiv Bxx + Lx + y = 0,$$

formulate the definitions (21.13), (21.14), and (21.15) of the Newton sequence $\{x_m\}$. If the solution of a linear equation and the inversion of a linear operator are regarded to be computationally equivalent, which formulation apparently requires the least computational effort?
3. Give the formulation (21.13) of Newton's method for the complex scalar equation

$$f(z) = 0$$

considered as a system of two real equations

$$u(x, y) = 0,$$

$$v(x, y) = 0,$$

where $z = x + iy$, $f(z) = u(x, y) + iv(x, y)$. *Hint.* Use the Cauchy-Riemann conditions [6] to simplify the expressions obtained.

4. In Problem 2, consider the abstract cubic equation

$$C(x) \equiv Txxx + Bxx + Lx + y = 0.$$

22. THE CONVERGENCE OF NEWTON'S METHOD

Attention will now be given to the conditions under which the Newton sequence $\{x_m\}$ defined in the previous section will converge to a solution $x = x^*$ of the nonlinear equation

$$P(x) = 0. \tag{22.1}$$

For some $x_0 \in X$, it will be assumed that the Newton sequence is given by

$$x_{m+1} = x_m - [P'(x_m)]^{-1} P(x_m), \qquad m = 0, 1, 2, \ldots. \tag{22.2}$$

Of course, an infinite Newton sequence will not necessarily be generated for every choice of the initial point x_0. For example, the operator $P'(x_k)$ may not have an inverse for some integer k. If $P(x_k) \neq 0$, then we call the finite sequence $\{x_0, x_1, \ldots, x_k\}$ obtained in this case *divergent owing to inversion failure at the kth step*. On the other hand, if $P(x_k) = 0$ [whether or not $P'(x_k)$ is invertible] and $P(x_m) \neq 0$ for any $m < k$, then $x^* = x_k$ is a solution of (22.1), and the sequence $\{x_0, x_1, \ldots, x_k\}$ is said to *converge to x^* at the kth step*.

If P and P' are defined on the entire space X, then $\{x_m\}$ will be an infinite sequence unless one of the two possibilities discussed above obtain. If the sequence $\{x_m\}$ is infinite, then it will either diverge or converge to some $x^* \in X$ in the usual sense. Under fairly mild restrictions the limit x^* of a convergent sequence $\{x_m\}$ will be a solution of (22.1). Suppose that

$$x^* = \lim_{m \to \infty} x_m. \tag{22.3}$$

The three following theorems give sufficient conditions for x^* to be a solution of (22.1).

Theorem 22.1. If P' is continuous at $x = x^*$, then

$$P(x^*) = 0. \tag{22.4}$$

PROOF. The elements of the sequence $\{x_m\}$ satisfy the equation

$$P'(x_m)(x_{m+1} - x_m) = -P(x_m). \tag{22.5}$$

Since the continuity of P at x^* follows from the continuity of P', taking the limit as $m \to \infty$ in (22.5) gives (22.4).

Theorem 22.2.   If

$$\|P'(x)\| \leq M \tag{22.6}$$

in some closed ball which contains $\{x_m\}$, then x^* is a solution of (22.1).

PROOF. Inequalities (22.6) and (20.45) *page 125* (see Corollary 20.3 to Taylor's Theorem 20.2) imply that

$$\lim_{m \to \infty} P(x_m) = P(x^*), \tag{22.7}$$

and since

page 125
$$\|P(x_m)\| \leq M \|x_{m+1} - x_m\| \tag{22.8}$$

by (20.45), (22.5), and (22.6), (22.4) is obtained by taking the limit as $m \to \infty$ in (22.8).

Theorem 22.3.   If

$$\|P''(x)\| \leq K \tag{22.9}$$

in some closed ball $\bar{U}(x_0, r)$, $0 < r < \infty$, which contains $\{x_m\}$, then x^* is a solution of (22.1).

PROOF.   Inequalities (22.9) and (20.45) imply that

$$\|P'(x) - P'(x_0)\| \leq K \|x - x_0\| \leq Kr \tag{22.10}$$

for all $x \in \bar{U}(x_0, r)$. Thus since

$$\|P'(x)\| \leq \|P'(x_0)\| + \|P'(x) - P'(x_0)\|, \tag{22.11}$$

the conditions of Theorem 22.2 hold with

$$M = \|P'(x_0)\| + Kr. \tag{22.12}$$

It should be noted that under the hypotheses of the foregoing theorems the convergence of the Newton sequence $\{x_m\}$ implies the *existence* of a solution $x = x^*$ of (22.1). Thus if the existence and convergence of the Newton sequence can be established, it is fairly easy to assure ourselves that the original equation (22.1) has a solution. For this purpose the basic tool used here will be the theorem on the convergence of Newton's method in Banach spaces as formulated by L. V. Kantorovič [3]. This theorem gives not only conditions for the existence of x^* but also information concerning the regions of existence and uniqueness of x^* and error bounds for the terms x_m of the Newton sequence as approximations to x^*. Thus this theorem provides a computational theory of (22.1).

To discuss the convergence of the Newton sequence $\{x_m\}$, starting from some point x_0, it is assumed first of all that $[P'(x_0)]^{-1}$ exists. This permits the calculation of the next point

$$x_1 = x_0 - [P'(x_0)]^{-1} P(x_0) \tag{22.13}$$

and constants B_0, η_0 such that

$$\|[P'(x_0)]^{-1}\| \leq B_0, \tag{22.14}$$

$$\|x_1 - x_0\| \leq \eta_0, \tag{22.15}$$

respectively.

Theorem 22.4 (*Kantorovič*).   If

$$\|P''(x)\| \leq K \tag{22.16}$$

in some closed ball $\bar{U}(x_0, r)$ and

$$h_0 = B_0 \eta_0 K \leq \tfrac{1}{2}, \tag{22.17}$$

then the Newton sequence (22.2), starting from x_0, will converge to a solution x^* of (22.1) which exists in $\bar{U}(x_0, r)$, provided that

$$r \geq r_0 = \frac{1 - \sqrt{1 - 2h_0}}{h_0} \eta_0. \tag{22.18}$$

PROOF.   This theorem is proved by mathematical induction, with the general inductive step being equivalent to the passage from x_0 to x_1. First of all, to continue to generate the Newton sequence at x_1, $[P'(x_1)]^{-1}$ must exist. From (22.16) and (20.45)

$$\|P'(x_1) - P'(x_0)\| \leq K \|x_1 - x_0\| = K\eta_0. \tag{22.19}$$

From (22.17)

$$\|P'(x_1) - P'(x_0)\| \leq K\eta_0 \leq \frac{1}{2B_0} < \frac{1}{\|[P'(x_0)]^{-1}\|}. \tag{22.20}$$

By Theorem 10.1, $[P'(x_1)]^{-1}$ exists, and

$$\|[P'(x_1)]^{-1}\| \leq \frac{\|[P'(x_0)]^{-1}\|}{1 - \|[P'(x_0)]^{-1}\| \cdot \|P'(x_1) - P'(x_0)\|} \tag{22.21}$$

or

$$\|[P'(x_1)]^{-1}\| \leq \frac{B_0}{1 - B_0\eta_0 K} = \frac{B_0}{1 - h_0} = B_1. \tag{22.22}$$

Consequently it is possible to calculate x_2. To estimate $\|x_2 - x_1\|$, note that

$$\|x_2 - x_1\| = \|[P'(x_1)]^{-1} P(x_1)\| \tag{22.23}$$

and

$$[P'(x_1)]^{-1} = \left(\sum_{n=0}^{\infty} \{I - [P'(x_0)]^{-1} P'(x_1)\}^n \right) [P'(x_0)]^{-1}$$

$$= \left(\sum_{n=0}^{\infty} \{[P'(x_0)]^{-1}[P'(x_0) - P'(x_1)]\}^n \right) [P'(x_0)]^{-1}. \tag{22.24}$$

Consequently

$$\|[P'(x_1)]^{-1} P(x_1)\| \leq \frac{1}{1 - B_0\eta_0 K} \|[P'(x_0)]^{-1} P(x_1)\| \qquad (22.25)$$

or

$$\|x_2 - x_1\| \leq \frac{1}{1 - h_0} \|[P'(x_0)]^{-1} P(x_1)\|. \qquad (22.26)$$

To estimate $\|[P'(x_0)]^{-1} P(x_1)\|$ consider the operator

$$F(x) = x - [P'(x_0)]^{-1} P(x). \qquad (22.27)$$

Obviously

$$F'(x_0) = I - [P'(x_0)]^{-1} P'(x_0) = 0. \qquad (22.28)$$

From (22.27)

$$[P'(x_0)]^{-1} P(x_1) = x_1 - F(x_1), \qquad (22.29)$$

and since

$$x_1 = F(x_0), \qquad (22.30)$$

(22.28) may be used to write (22.29) as

$$[P'(x_0)]^{-1} P(x_1) = -[F(x_1) - F(x_0) - F'(x_0)(x_1 - x_0)], \qquad (22.31)$$

so that by (20.45)

$$\|[P'(x_0)]^{-1} P(x_1)\| \leq \sup_{\bar{x} \in L(x_0, x_1)} \|F''(\bar{x})\| \frac{\|x_1 - x_0\|^2}{2}. \qquad (22.32)$$

From (22.27)

$$F''(x) = -[P'(x_0)]^{-1} P''(x). \qquad (22.33)$$

Since $x_1 \in \bar{U}(x_0, r)$ if (22.28) holds, then *Why does $x_1 \in \bar{U}(x_0, r)$*

$$\|F''(x)\| \leq B_0 K \qquad (22.34)$$

on $L(x_0, x_1)$. From (22.32)

$$\|[P'(x_0)]^{-1} P(x_1)\| \leq \frac{B_0 K \eta_0^2}{2} = \frac{h_0\eta_0}{2}, \qquad (22.35)$$

and from (22.26)

$$\|x_2 - x_1\| \leq \frac{1}{2} \frac{h_0}{1 - h_0} \eta_0 = \eta_1 \qquad (22.36)$$

and obviously $\eta_1 \leq \frac{1}{2}\eta_0$. The constant

$$h_1 = B_1\eta_1 K = \frac{B_0}{1 - h_0} \frac{1}{2} \frac{h_0}{1 - h_0} \eta_0 K$$

$$= \frac{1}{2} \frac{h_0^2}{(1 - h_0)^2} \leq 2h_0^2 \leq \tfrac{1}{2}, \qquad (22.37)$$

so that (22.17) is satisfied if x_0 is replaced by x_1. Now if for

$$r_1 = \frac{1 - \sqrt{1 - 2h_1}}{h_1} \eta_1 \tag{22.38}$$

it can be shown that

$$\bar{U}(x_1, r_1) \subset \bar{U}(x_0, r_0), \tag{22.39}$$

then condition (22.16) will hold with x_0 replaced by x_1. By direct substitution

$$\sqrt{1 - 2h_1} = \left(1 - \frac{h_0^2}{(1 - h_0)^2}\right)^{1/2} = \frac{1}{1 - h_0} \sqrt{1 - 2h_0} \tag{22.40}$$

and

$$r_1 = \frac{1 - (1 - h_0)^{-1}\sqrt{1 - 2h_0}}{\frac{1}{2}[h_0^2/(1 - h_0)^2]} \frac{1}{2} \frac{h_0}{1 - h_0} \eta_0$$

$$= \left(\frac{1 - \sqrt{1 - 2h_0}}{h_0} - 1\right)\eta_0 = r_0 - \eta_0. \tag{22.41}$$

Consequently, if $x \in \bar{U}(x_1, r_1)$, then $\|x - x_1\| \leq r_1 = r_0 - \eta_0$ and

$$\|x - x_0\| \leq \|x - x_1\| + \|x_1 - x_0\| \leq (r_0 - \eta_0) + \eta_0 = r_0, \tag{22.42}$$

so that $x \in \bar{U}(x_0, r_0)$, which establishes (22.39).

It follows by mathematical induction that the Newton process (22.2) generates an infinite sequence $\{x_m\}$, starting from an x_0 at which the hypotheses of the theorem are satisfied. It remains to be shown that this sequence converges to a solution x^* of (22.1). Along with $\{x_m\}$ the sequences of numbers $\{B_m\}$, $\{\eta_m\}$, and $\{h_m\}$ defined by

$$B_m = \frac{B_{m-1}}{1 - h_{m-1}}, \tag{22.43}$$

$$\eta_m = \frac{1}{2} \frac{h_{m-1}\eta_{m-1}}{1 - h_{m-1}}, \tag{22.44}$$

$$h_m = \frac{1}{2} \frac{h_{m-1}^2}{(1 - h_{m-1})^2}, \tag{22.45}$$

respectively, are obtained for $m = 1, 2, \ldots$. We have

$$h_m \leq 2h_{m-1}^2 \leq \tfrac{1}{2}(2h_{m-2})^4 \leq \cdots \leq \tfrac{1}{2}(2h_0)^{2^m} \tag{22.46}$$

from (22.37) and

$$\eta_m = \frac{1}{2} \frac{h_{m-1}\eta_{m-1}}{1 - h_{m-1}} \leq h_{m-1}\eta_{m-1} \leq h_{m-1}h_{m-2}\eta_{m-2} \leq \cdots$$

$$\leq \frac{1}{2^m} (2h_0)^{2^{m-1}}(2h_0)^{2^{m-2}} \cdots (2h_0)\eta_0 = \frac{1}{2^m} (2h_0)^{2^m-1}\eta_0. \tag{22.47}$$

For any positive integer p, $x_{m+p} \in \bar{U}(x_m, r_m)$, so that

$$\|x_{m+p} - x_m\| \leq r_m = \frac{1 - \sqrt{1 - 2h_m}}{h_m} \eta_m \leq 2\eta_m, \qquad (22.48)$$

and thus from (22.47)

$$\|x_{m+p} - x_m\| \leq \frac{1}{2^{m-1}} (2h_0)^{2^m - 1} \eta_0. \qquad (22.49)$$

Therefore $\{x_m\}$ is a Cauchy sequence which has a limit $x^* \in \bar{U}(x_0, r_0)$, and x^* is a solution of (22.1) by Theorem 22.3. This completes the proof of Theorem 22.4.

An error estimate may be obtained immediately from the fact that

$$\{x_{m+1}, x_{m+2}, \ldots\} \subset \bar{U}(x_m, r_m). \qquad (22.50)$$

Theorem 22.5.   If the conditions of Theorem 22.4 are satisfied, then

$$\|x^* - x_m\| \leq \frac{1}{2^{m-1}} (2h_0)^{2^m - 1} \eta_0. \qquad (22.51)$$

PROOF.   It follows from the convergence of the Newton sequence $\{x_m\}$ and (22.50) that $x^* \in \bar{U}(x_m, r_m)$, so that

$$\|x^* - x_m\| \leq r_m \leq 2\eta_m, \qquad (22.52)$$

as in (22.48), and (22.51) follows from (22.47).

Inequality (22.51) indicates that for $h_0 < \frac{1}{2}$ the Newton sequence converges more rapidly than the sequence $\{x_m\}$ for the solution of the linear equation $Lx = y$ given by (9.48). The rapidity of convergence of the Newton process for small values of h_0 is clear if we introduce the *improvement factor* [7]

$$\theta_m = \frac{(2h_0)^{2^m}}{2^m(1 - \sqrt{1 - 2h_0})}, \qquad (22.53)$$

which is a function of h_0 only, and

$$\theta_m \geq \frac{r_m}{r_0} \qquad (22.54)$$

from (22.18), (22.47), and (22.48). The values of the minimum N of m such that

$$\theta_N \leq 10^{-k} \qquad (22.55)$$

are given in Table 22.1 for $k = 1(1)20$ and $h_0 = 0.01(0.01)0.50$.

The question of uniqueness may be approached essentially on the basis that (22.16) holds in some ball which contains $\bar{U}(x_0, r_0)$. It is convenient to consider the case $h_0 = \frac{1}{2}$ separately.

Theorem 22.6. If $h_0 < \frac{1}{2}$ and (22.16) holds in the open ball $U(x_0, \tilde{r}_0)$, where

$$\tilde{r}_0 = \frac{1 + \sqrt{1 - 2h_0}}{h_0} \eta_0, \qquad (22.56)$$

then (22.1) has a solution x^* which is unique in $U(x_0, \tilde{r}_0)$.

PROOF.   Since

$$\bar{U}(x_0, r_0) \subset U(x_0, \tilde{r}_0), \qquad (22.57)$$

it follows from Theorem 22.4 that (22.1) has a solution $x^* \in \bar{U}(x_0, r_0)$ to which the Newton sequence $\{x_m\}$ converges. Suppose that $\tilde{x}$ is any solution of the equation $P(x) = 0$ in $U(x_0, \tilde{r}_0)$. It follows from (22.27) and the fact that $P(\tilde{x}) = 0$ that

$$\tilde{x} = F(\tilde{x}). \qquad (22.58)$$

Thus

$$\|\tilde{x} - x_1\| = \|F(\tilde{x}) - F(x_0)\| = \|F(\tilde{x}) - F(x_0) - F'(x_0)(\tilde{x} - x_0)\|, \qquad (22.59)$$

since $F'(x_0) = 0$ and thus

$$\|\tilde{x} - x_1\| \le \tfrac{1}{2} B_0 K \|\tilde{x} - x_0\|^2 \qquad (22.60)$$

by (20.45) and (22.34).

Assuming that

$$\|\tilde{x} - x_0\| = \theta \tilde{r}_0 = \theta \frac{1 + \sqrt{1 - 2h_0}}{h_0} \eta_0, \qquad 0 < \theta < 1, \qquad (22.61)$$

we note that

$$
\begin{aligned}
\tfrac{1}{2} B_0 K \tilde{r}_0^{\,2} &= \tfrac{1}{2} h_0 \left(\frac{1 + \sqrt{1 - 2h_0}}{h_0} \right)^2 \eta_0 \\
&= \left(\frac{1 + \sqrt{1 - 2h_0}}{h_0} \right)^2 (1 - h_0) \eta_1 \\
&= \frac{(1 - h_0)[1 + 2\sqrt{1 - 2h_0} + (1 - 2h_0)]}{h_0^2} \eta_1 \\
&= \frac{2(1 - h_0)}{h_0^2} (1 - h_0 + \sqrt{1 - 2h_0}) \eta_1 \\
&= \frac{2(1 - h_0)^2}{h_0^2} (1 + \sqrt{1 - 2h_1}) \eta_1 \\
&= \frac{1 + \sqrt{1 - 2h_1}}{h_1} \eta_1 = \tilde{r}_1 \qquad (22.62)
\end{aligned}
$$

TABLE 22.1

h_0 \ k	1	2	3	4	5	6	7	8	9	10	11	12	13	14	15	16	17	18	19	20
0.010	1	2	2	2	2	3	3	3	3	3	3	3	4	4	4	4	4	4	4	4
0.020	1	2	2	2	3	3	3	3	3	3	3	4	4	4	4	4	4	4	4	4
0.030	1	2	2	3	3	3	3	3	3	4	4	4	4	4	4	4	4	4	4	5
0.040	1	2	2	3	3	3	3	3	4	4	4	4	4	4	4	4	4	5	5	5
0.050	1	2	2	3	3	3	3	4	4	4	4	4	4	4	5	5	5	5	5	5
0.060	2	2	2	3	3	3	3	4	4	4	4	4	4	5	5	5	5	5	5	5
0.070	2	2	3	3	3	3	4	4	4	4	4	4	5	5	5	5	5	5	5	5
0.080	2	2	3	3	3	4	4	4	4	4	4	4	5	5	5	5	5	5	5	5
0.090	2	2	3	3	3	4	4	4	4	4	5	5	5	5	5	5	5	5	5	5
0.100	2	2	3	3	4	4	4	4	4	4	5	5	5	5	5	5	5	5	5	5
0.110	2	2	3	3	4	4	4	4	4	5	5	5	5	5	5	5	5	5	5	5
0.120	2	2	3	3	4	4	4	4	4	5	5	5	5	5	5	5	5	5	5	5
0.130	2	3	3	3	4	4	4	4	5	5	5	5	5	5	5	5	5	5	6	6
0.140	2	3	3	3	4	4	4	4	5	5	5	5	5	5	5	5	5	6	6	6
0.150	2	3	3	4	4	4	4	5	5	5	5	5	5	5	6	6	6	6	6	6
0.160	2	3	3	4	4	4	4	5	5	5	5	5	5	6	6	6	6	6	6	6
0.170	2	3	3	4	4	4	4	5	5	5	5	5	5	6	6	6	6	6	6	6
0.180	2	3	3	4	4	4	5	5	5	5	5	5	6	6	6	6	6	6	6	6
0.190	2	3	3	4	4	4	5	5	5	5	5	5	6	6	6	6	6	6	6	6
0.200	2	3	3	4	4	4	5	5	5	5	5	5	6	6	6	6	6	6	6	6
0.210	2	3	3	4	4	4	5	5	5	5	5	5	6	6	6	6	6	6	6	6
0.220	2	3	3	4	4	4	5	5	5	5	5	5	6	6	6	6	6	6	6	6

	7	10	14	17	20	24	27	30	34	37	40	44	47	50	54	57	60	64	67
0.230	2	3	4	4	4	5	5	5	5	5	6	6	6	6	6	6	6	6	6
0.240	2	3	4	4	5	5	5	5	5	5	6	6	6	6	6	6	6	6	6
0.250	2	3	4	4	5	5	5	5	5	6	6	6	6	6	6	6	6	6	6
0.260	2	3	4	4	5	5	5	5	6	6	6	6	6	6	6	6	6	6	7
0.270	2	3	4	5	5	5	5	6	6	6	6	6	6	6	6	6	6	7	7
0.280	2	3	4	5	5	5	6	6	6	6	6	6	6	6	6	6	7	7	7
0.290	2	3	4	5	5	6	6	6	6	6	6	6	6	6	6	7	7	7	7
0.300	2	4	5	5	6	6	6	6	6	6	6	6	6	7	7	7	7	7	7
0.310	3	4	5	5	6	6	6	6	6	6	6	6	7	7	7	7	7	7	7
0.320	3	4	5	5	6	6	6	6	6	6	6	7	7	7	7	7	7	7	7
0.330	3	4	5	5	6	6	6	6	6	6	7	7	7	7	7	7	7	7	7
0.340	3	4	5	5	6	6	6	6	6	7	7	7	7	7	7	7	7	7	7
0.350	3	4	5	5	6	6	6	6	7	7	7	7	7	7	7	7	7	7	7
0.360	3	4	5	5	6	6	6	7	7	7	7	7	7	7	7	7	7	7	7
0.370	3	4	5	6	6	6	7	7	7	7	7	7	7	7	7	7	7	7	7
0.380	3	4	5	6	6	6	7	7	7	7	7	7	7	7	7	7	7	7	8
0.390	3	4	5	6	6	6	7	7	7	7	7	7	7	7	7	7	7	8	8
0.400	3	4	6	6	6	6	7	7	7	7	7	7	7	7	8	8	8	8	9
0.410	3	5	6	6	6	7	7	7	7	7	7	7	8	8	8	8	8	8	9
0.420	3	5	6	6	6	7	7	7	7	7	8	8	8	8	8	8	8	9	9
0.430	3	5	6	6	7	7	7	7	7	8	8	8	8	8	8	8	9	9	9
0.440	3	5	6	6	7	7	7	7	8	8	8	8	8	8	8	9	9	9	9
0.450	5	5	6	6	7	7	8	8	8	8	8	9	9	9	9	9	9	9	9
0.460	5	5	6	7	7	8	8	8	9	9	9	9	9	9	9	9	9	9	9
0.470	5	6	7	7	8	8	9	9	9	9	10	10	10	10	10	10	10	10	10
0.480	6	6	7	8	8	9	10	10	10	10	11	11	11	11	11	11	10	10	10
0.490	6	7	8	9	9	9	10	10	11	11	11	11	11	11	11	11	11	11	11
0.500	7	10	14	17	20	24	27	30	34	37	40	44	47	50	54	57	60	64	67

141

by (22.36), (22.37), and (22.40). Consequently from (22.60) through (22.62)

$$\|\tilde{x} - x_n\| \leq \theta^{2^n} \tilde{r}_n = \theta^{2^n} \frac{1 + \sqrt{1 - 2h_n}}{h_n} \eta_n. \tag{22.63}$$

Since

$$\tilde{r}_n = \frac{1 + \sqrt{1 - 2h_n}}{h_n} \eta_n \leq \frac{2\eta_n}{h_n} = \frac{2}{B_n K} \tag{22.64}$$

and $B_n > B_0$ by (22.45) and induction, (22.63) and (22.64) give

$$\|\tilde{x} - x_n\| \leq \theta^{2^n} \frac{2}{B_0 K}. \tag{22.65}$$

Since $0 < \theta < 1$, $\lim_{m \to \infty} x_m = \tilde{x}$. Since the limit of the Newton sequence $\{x_m\}$ is unique, $\tilde{x} = x^*$, so that x^* is the only solution of (22.1) in the ball $U(x_0, \tilde{r}_0)$. The theorem is proved.

Theorem 22.7. If the conditions of Theorem 22.4 are satisfied and $h_0 = \frac{1}{2}$, then there exists a unique solution x^* of (22.1) in the closed ball

$$\bar{U}(x_0, r_0) = \bar{U}(x_0, 2\eta_0).$$

PROOF.   The analysis in the proof of Theorem 22.6 holds, except that it is possible that $\theta = 1$ in (22.61). This would give in place of (22.63) and (22.64),

$$\|\tilde{x} - x_n\| \leq r_n \leq \frac{2}{B_n K}. \tag{22.66}$$

In the case that $h_0 = \frac{1}{2}$, it follows from (22.43) that

$$B_1 = 2B_0, \quad B_2 = 2B_1 = 4B_0, \ldots, \quad B_n = 2^n B_0, \tag{22.67}$$

so that from (22.66)

$$\|\tilde{x} - x_n\| \leq \frac{1}{2^{n-1}} \frac{1}{B_0 K}. \tag{22.68}$$

Here $\lim_{m \to \infty} x_m = \tilde{x}$ also and thus $\tilde{x} = x^*$, the unique solution of (22.1) in the ball $\bar{U}(x_0, r_0) = \bar{U}(x_0, 2\eta_0)$. The theorem is proved.

From the standpoint of actual practice, the choice of x_0 determines at once whether or not $[P'(x_0)]^{-1}$ exists. If $[P'(x_0)]^{-1}$, then the numbers B_0 and η_0 may be obtained at once. When it is not feasible to calculate the point x_1, the estimate

$$\|x_1 - x_0\| \leq B_0 \|P(x_0)\| \tag{22.69}$$

may be used to obtain a value η_0 which satisfies (22.15). Once the numbers B_0, η_0 are determined, it is only necessary to know something about the behavior of $\|P''(x)\|$ in a neighborhood of x_0 in order to verify the satisfaction of the hypotheses of Theorems 22.4 through 22.7. For a number of types of equations to be considered later, it is not difficult to obtain estimates for $\|P''(x)\|$.

It is possible to make a graphical interpretation of the conditions for the existence and uniqueness of a solution x^* of (22.1) as given by the theorems of this section in terms of the "dimensionless" quantities h_0 and

$$s_0 = \frac{r}{\eta_0} \le \frac{\|x - x_0\|}{\|x_1 - x_0\|}. \tag{22.70}$$

From (22.18)

$$h_0 = \frac{2(s_0 - 1)}{s_0{}^2}. \tag{22.71}$$

If it is now known that

$$\|P''(x)\| \le K(s_0), \tag{22.72}$$

where $K(s_0)$ is a nonnegative, monotone increasing function such that

$$K(0) = \|P''(x_0)\|, \tag{22.73}$$

then Theorem 22.4 guarantees the existence of x^* if the curve

$$h_0 = B_0\eta_0 K(s_0) \tag{22.74}$$

intersects the curve (22.71) in the region $1 \le s_0 \le 2$ (see Figure 22.1). The intersection point with the minimum abscissa $s_0{}^E$ determines the radius

$$r_0{}^E = s_0{}^E \eta_0 \tag{22.75}$$

of the closed ball $\bar{U}(x_0, r_0{}^E)$ in which the existence of x^* is guaranteed by Theorem 22.4. The intersection point with maximum abscissa $s_0{}^U$ determines the radius

$$r_0{}^U = s_0{}^U \eta_0 \tag{22.76}$$

of the open ball $U(x_0, r_0{}^U)$ in which Theorem 22.6 guarantees that x^* is unique, provided that

$$s_0{}^U > 2. \tag{22.77}$$

If there is only the single intersection point

$$s_0 = 2, \qquad h_0 = \tfrac{1}{2}, \tag{22.78}$$

then Theorem 22.7 applies, and x^* exists and is unique in the closed ball $\bar{U}(x_0, 2\eta_0)$.

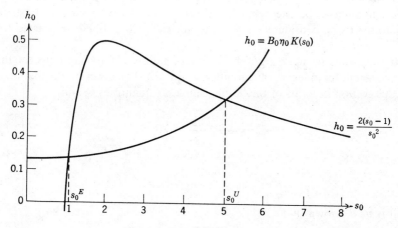

Figure 22.1 $h_0 = 2(s_0 - 1)/s_0^2$.

EXERCISES 22

1. Describe the Newton sequences for the real scalar operators $P(x) = x^2 - 1$ and $P(x) = x^2 + 1$ for various real x_0.

2. For the real scalar operator

$$P(x) = x^3 - 4x,$$

   show that it is possible to find points x_0 such that the Newton sequence will converge to $x^* = 2$ (or $x^* = -2$) in k steps, for $k = 1, 2, \ldots$. Also show that it is possible to find points x_0 such that the Newton sequence diverges in k steps owing to inversion failure, once again for $k = 1, 2, \ldots$. *Hint.* Interpret geometrically.

3. Discuss the application of the theory of the convergence of Newton's method to a solution of the real quadratic equation

$$P(x) \equiv ax^2 + bx + c = 0$$

   for real x_0.

4. After the calculation of x_1 and $P(x_1)$, show that the error estimate

$$\|x^* - x_1\| \leq \frac{2}{1 - h_0 + \sqrt{1 - 2h_0}} B_0 \|P(x_1)\|$$

   holds if $h_0 \leq \tfrac{1}{2}$.

5. Show on the basis of the real quadratic equation

$$P(x) \equiv \tfrac{1}{2}Kx^2 - \frac{1}{B_0}x + \frac{\eta_0}{B_0} = 0$$

   that all of the estimates in Theorems 22.4–22.7 may be attained.

6. Apply the theorems on the convergence of Newton's method to the sequence (21.22) for the calculation of $\sqrt[k]{N}$, assuming that N is real, positive, and that $x_0 = 1$. Draw figures corresponding to Figure 22.1 for $k = 2, 3, 4$, and $N = \frac{1}{3}, 3$.

23. APPLICATION OF NEWTON'S METHOD TO THE SOLUTION OF FINITE SYSTEMS OF EQUATIONS

From the standpoint of actual computation, the most important application of Newton's method is to the solution of finite systems of nonlinear equations. Such systems may arise directly or result from approximation of nonlinear integral or differential equations by the application of an appropriate discretization technique.

For our purposes it is sufficiently general to consider (21.1) in $X = R^n$ as a system of the form

$$P_1(\xi_1, \xi_2, \ldots, \xi_n) = 0,$$

$$P_2(\xi_1, \xi_2, \ldots, \xi_n) = 0,$$

$$\vdots \tag{23.1}$$

$$P_n(\xi_1, \xi_2, \ldots, \xi_n) = 0,$$

or in vector notation,

$$P(x) = (P_1(x), P_2(x), \ldots, P_n(x)) = 0, \tag{23.2}$$

where

$$x = (\xi_1, \xi_2, \ldots, \xi_n) \tag{23.3}$$

and each $P_i(\xi_1, \xi_2, \ldots, \xi_n)$, $i = 1, 2, \ldots, n$, is a real function of the real variables $\xi_1, \xi_2, \ldots, \xi_n$.

It would also be possible to consider finite systems of equations

$$F_1(\zeta_1, \zeta_2, \ldots, \zeta_m) = 0,$$

$$F_2(\zeta_1, \zeta_2, \ldots, \zeta_m) = 0,$$

$$\vdots \tag{23.4}$$

$$F_m(\zeta_1, \zeta_2, \ldots, \zeta_m) = 0,$$

where the F_k, $k = 1, 2, \ldots, m$, are complex-valued functions of the complex variables $\zeta_1, \zeta_2, \ldots, \zeta_m$ as an equation of the form $F(z) = 0$ in the space $X = C^m$ of the vectors $z = (\zeta_1, \zeta_2, \ldots, \zeta_m)$. However, since most computation is done with real arithmetic, it is customary to write (23.4) in terms of real and imaginary parts by setting

$$\zeta_k = \xi_{2k-1} + i\xi_{2k} \qquad (23.5)$$

and

$$F_k(\zeta_1, \zeta_2, \ldots, \zeta_m) = P_{2k-1}(\xi_1, \xi_2, \ldots, \xi_{2m}) + iP_{2k}(\xi_1, \xi_2, \ldots, \xi_{2m}),$$
$$k = 1, 2, \ldots, m. \qquad (23.6)$$

A system of real equations of the form (23.1) with $n = 2m$ is obtained by setting the real and imaginary parts of the functions (23.6) equal to zero. [In (23.5) and (23.6), i denotes the imaginary unit, $i \equiv \sqrt{-1}$.] Thus the equation $F(z) = 0$ in $X = C^m$ may be regarded as being an equation of the form $P(x) = 0$ in $X = R^{2m}$.

Attention will now be given to a detailed description of the application of Newton's method to the solution of the system (23.1), including convergence and error analysis. First, in order to generate the Newton sequence, it is necessary to choose an initial approximation

$$x_0 = (\xi_1^{(0)}, \xi_2^{(0)}, \ldots, \xi_n^{(0)}) \qquad (23.7)$$

to a solution $x^* = (\xi_1^*, \xi_2^*, \ldots, \xi_n^*)$ of (23.1). Of course, it is not required to know in advance that a solution x^* exists in the vicinity of x_0 or some subsequent iterate, as the existence of x^* can be established on the basis of Theorem 22.4. In addition to the n functions

$$P_i(x) = P_i(\xi_1, \xi_2, \ldots, \xi_n), \qquad i = 1, 2, \ldots, n, \qquad (23.8)$$

we must also be able to evaluate the n^2 partial derivatives

$$P'_{ij}(x) = \frac{\partial P_i(\xi_1, \xi_2, \ldots, \xi_n)}{\partial \xi_j}, \qquad i, j = 1, 2, \ldots, n \quad \text{at} \quad x = x_0. \qquad (23.9)$$

It is assumed, of course, that the functions (23.8) and (23.9) can be evaluated by a finite number of arithmetic operations, at least to any desired degree of accuracy. For the present, it will be supposed that all calculations are performed exactly, the modifications in the theory for the case of approximate evaluation being indicated later. As a result of these calculations, we have the vector

$$P(x_0) = (P_1(x_0), P_2(x_0), \ldots, P_n(x_0)) \qquad (23.10)$$

and the matrix

$$P'(x_0) = (P'_{ij}(x_0)). \qquad (23.11)$$

The next step in the generation of the Newton sequence, the computation of

$$x_1 = (\xi_1^{(1)}, \xi_2^{(1)}, \ldots, \xi_n^{(1)}) \qquad (23.12)$$

will require the inversion of the $n \times n$ matrix (23.11) if (21.13) is used, or the solution of n linear algebraic equations in n unknowns with the coefficient matrix (23.11) if (21.14) or (21.15) is applied. If

$$G_0 = (g_{ij}^{(0)}) = [P'(x_0)]^{-1}, \tag{23.13}$$

then (21.13) becomes

$$\xi_i^{(1)} = \xi_i^{(0)} - \sum_{j=1}^{n} g_{ij}^{(0)} P_j(\xi_1^{(0)}, \xi_2^{(0)}, \ldots, \xi_n^{(0)}), \qquad i = 1, 2, \ldots, n. \tag{23.14}$$

For

$$p_{ij}^{(0)} = P'_{ij}(x_0), \qquad i, j = 1, 2, \ldots, n, \tag{23.15}$$

(21.14) becomes the system

$$\sum_{j=1}^{n} p_{ij}^{(0)} \xi_j^{(1)} = \sum_{j=1}^{n} p_{ij}^{(0)} \xi_j^{(0)} - P_i(\xi_1^{(0)}, \xi_2^{(0)}, \ldots, \xi_n^{(0)}), \qquad i = 1, 2, \ldots, n. \tag{23.16}$$

Similarly, we may set

$$\Delta x_0 = (\Delta \xi_1^{(0)}, \Delta \xi_2^{(0)}, \ldots, \Delta \xi_n^{(0)}) \tag{23.17}$$

and obtain the system (21.15) in the form

$$\sum_{j=1}^{n} p_{ij}^{(0)} \Delta \xi_j^{(0)} = -P_i(\xi_1^{(0)}, \xi_2^{(0)}, \ldots, \xi_n^{(0)}), \qquad i = 1, 2, \ldots, n, \tag{23.18}$$

from the solution of which we have

$$\xi_i^{(1)} = \xi_i^{(0)} + \Delta \xi_i^{(0)}, \qquad i = 1, 2, \ldots, n. \tag{23.19}$$

At first glance it appears that the application of (23.16) is more complicated than the use of (23.18) and (23.19). For many operators P, however, the right-hand side $P'(x)x - P(x)$ of (23.16) may be relatively simple in form when calculated explicitly; for example, suppose that the operator P in the space R^3 of real vectors $x = (\xi, \eta, \zeta)$ is defined by

$$P\begin{pmatrix} \xi \\ \eta \\ \zeta \end{pmatrix} = \begin{pmatrix} 16\xi^4 + 16\eta^4 + \zeta^4 - 16 \\ \xi^2 + \eta^2 + \zeta^2 - 3 \\ \xi^3 - \eta \end{pmatrix} \tag{23.20}$$

Here $P'(x)x - P(x)$ has the form

$$\left[P'\begin{pmatrix} \xi \\ \eta \\ \zeta \end{pmatrix} \right]\begin{pmatrix} \xi \\ \eta \\ \zeta \end{pmatrix} - P\begin{pmatrix} \xi \\ \eta \\ \zeta \end{pmatrix} = \begin{pmatrix} 48\xi^4 + 48\eta^4 + 3\zeta^4 + 16 \\ \xi^2 + \eta^2 + \zeta^2 + 3 \\ 2\xi^3 \end{pmatrix}, \tag{23.21}$$

which is essentially as easy to evaluate as $P(x)$.

In any case, the calculation of x_1 will require the inversion of a matrix or the solution of a linear algebraic system, both of which are fairly well understood from a computational standpoint at the present time. Most computing centers have standard library programs which will ordinarily produce the desired results quickly and accurately, at least for systems of moderate size ($n \leq 100$). Such programs are usually based on some variant of gaussian elimination [8,9]. For large systems, which usually have some special structure, such as those which result from discretization of boundary-value problems for partial differential equations, iterative methods are frequently used [10,11].

For computational purposes the direct solution of a linear system is considered preferable to obtaining the same result by inverting the coefficient matrix followed by operation on the right-hand side [12]. This is partially offset by the fact that it is convenient to have the matrix $G_0 = [P'(x_0)]^{-1}$ in order to calculate the number B_0 needed for the convergence and error analysis on the basis of Theorem 22.4.

The continued generation of the Newton sequence can be accomplished by performing the same operations previously described with the value of the vector x_0 replaced by the value of x_1, and so on. This will be possible as long as the matrix $P'(x_m)$ remains nonsingular.

The second part of the application of Newton's method to the system (23.1) is the convergence and error analysis, which can be carried out along with the formation of the Newton sequence. After the calculation of x_1, we immediately obtain

$$\eta_0 = \|x_1 - x_0\|. \tag{23.22}$$

To be perfectly specific suppose that $x = R_\infty^m$, and thus,

$$\eta_0 = \max_{(i)} |\xi_i^{(1)} - \xi_i^{(0)}|. \tag{23.23}$$

(Of course, if preferable, any other norm could be used for R^m.) If x_1 has been calculated by (23.14), so that $G_0 = [P'(x_0)]^{-1}$ is known, then we can find

$$B_0 = \|G_0\| = \max_{(i)} \sum_{j=1}^{n} |g_{ij}^{(0)}| \tag{23.24}$$

immediately. The advantage to programming the computation to use (23.14) is thus that a separate calculation of the quantities needed for (23.24) is unnecessary; for example, consider the operator $P(x)$ in R_∞^3 defined by (23.20), and suppose that

$$x_0 = (1, 1, 1) \tag{23.25}$$

is taken to be an initial approximation to a solution $x = x^*$ of the equation

$P(x) = 0$. Since

$$P'\begin{pmatrix} \xi \\ \eta \\ \zeta \end{pmatrix} = \begin{pmatrix} 64\xi^3 & 64\eta^3 & 4\zeta^3 \\ 2\xi & 2\eta & 2\zeta \\ 3\xi^2 & -1 & 0 \end{pmatrix}, \tag{23.26}$$

it follows that

$$P'(x_0) = \begin{pmatrix} 64 & 64 & 4 \\ 2 & 2 & 2 \\ 3 & -1 & 0 \end{pmatrix}, \tag{23.27}$$

$$G_0 = [P'(x_0)]^{-1} = \frac{1}{240}\begin{pmatrix} 1 & -2 & 60 \\ 3 & -6 & -60 \\ -4 & 128 & 0 \end{pmatrix}, \tag{23.28}$$

and

$$x_1 = x_0 - G_0 P(x_0) = (\tfrac{223}{240}, \tfrac{63}{80}, \tfrac{79}{60}). \tag{23.29}$$

It follows at once that

$$\eta_0 = \tfrac{19}{60}, \qquad B_0 = \tfrac{11}{20} \tag{23.30}$$

for this example.

The next step in the application of Theorem 22.4 consists of determining a suitable upper bound $K \geq \|P''(x)\|$ for the second derivative in a sufficiently large ball. From (22.17), (22.18), and (23.30), if

$$K \leq \tfrac{600}{209} = 3.20855 \cdots \tag{23.31}$$

in a ball $\bar{U}(x_0, r)$, with

$$r \geq \frac{20(1 - \sqrt{1 - \tfrac{209}{600}K})}{11K}, \tag{23.32}$$

then Theorem 22.4 guarantees the existence in $\bar{U}(x_0, r)$ of a solution $x = x^*$ of the equation $P(x) = 0$, and the convergence of the Newton sequence to x^*.

In general, the second derivative of the vector operator

$$P(x) = (P_1(x), P_2(x), \ldots, P_n(x))$$

is the bilinear operator in R_∞^n with coefficients

$$P_{ijk}''(x) = \frac{\partial^2 P_i(\xi_1, \xi_2, \ldots, \xi_n)}{\partial \xi_j \partial \xi_k}, \qquad i, j, k = 1, 2, \ldots, n. \tag{23.33}$$

At first glance, it appears that the formation of the second derivative requires the calculation of n^3 partial derivatives. However, this is reduced by the symmetry relationships for twice differentiable operators, namely,

$$\frac{\partial^2 P_i(\xi_1, \xi_2, \ldots, \xi_n)}{\partial \xi_j \partial \xi_k} = \frac{\partial^2 P_i(\xi_1, \xi_2, \ldots, \xi_n)}{\partial \xi_k \partial \xi_j}, \qquad i, j, k = 1, 2, \ldots, n.$$

(23.34)

In $\bar{U}(x_0, r)$,

$$\|P''(x)\| \leq \sup_{x \in \bar{U}(x_0, r)} \left\{ \max_{(i)} \sum_{\substack{j=1 \\ k=1}}^{n} |P''_{ijk}(x)| \right\} = K(r).$$

(23.35)

From an expression of the type (23.35) it may be possible to derive an expression for $K(r)$ [or $K(s_0)$, $s_0 = r/n_0$] to use (22.17), (22.18), or the graphical method of Figure 22.1 to determine the existence of x^* and the convergence of Newton's method. For $P(x)$ defined by (23.20)

$$P''\begin{pmatrix} \xi \\ \eta \\ \zeta \end{pmatrix} = \begin{pmatrix} 192\xi^2 & 0 & 0 & 0 & 192\eta^2 & 0 & 0 & 0 & 12\zeta^2 \\ 2 & 0 & 0 & 0 & 2 & 0 & 0 & 0 & 2 \\ 6\xi & 0 & 0 & 0 & 0 & 0 & 0 & 0 & 0 \end{pmatrix}, \quad (23.36)$$

the notation being an extension of (17.33) to R_∞^3. For $x_0 = (1, 1, 1)$, it follows for $x \in \bar{U}(x_0, r)$ that

$$\max \{|\xi|, |\eta|, |\zeta|\} \leq 1 + r. \tag{23.37}$$

Thus from (23.35) and (23.36) we have

$$\|P''(x)\| \leq 396(1 + r)^2 \tag{23.38}$$

for $x \in \bar{U}(x_0, r)$. It follows from (23.38) that (23.31) is not satisfied at x_0, and so the convergence of Newton's method cannot be guaranteed. In the next section it will be shown that Newton's method actually converges in this case, with the hypotheses of Theorem 22.4 satisfied by a subsequent iterate.

The application of (23.35) is often not as difficult as we might suppose at first sight; for example, if $P(x)$ is a quadratic operator of the form

$$P(x) = y + Lx + Bxx, \tag{23.39}$$

where B is a symmetric bilinear operator, then

$$P''(x) = 2B \tag{23.40}$$

and

$$\dot{K} = 2\|B\|, \tag{23.41}$$

a constant. An example of this is furnished by the systems

$$\xi_i = 1 + \tfrac{1}{2}\varpi_0 \xi_i \sum_{j=1}^{9} b_{ij}\xi_j, \qquad i = 1, 2, \ldots, 9, \quad \varpi_0 = 0.1, 0.2, \ldots, 1.0,$$

(23.42)

which were considered in Section 13 as approximations to corresponding nonlinear integral equations. The coefficients b_{ij}, $i, j = 1, 2, \ldots, 9$ are given in Table 13.2. The systems (23.42) are of the form (23.39) with

$$y = (1, 1, \ldots, 1),$$
$$L = -I = -(\delta_{ij}), \tag{23.43}$$
$$B = (\beta_{ijk}) = \tfrac{1}{4}\varpi_0(b_{ij}\delta_{ik} + b_{ik}\delta_{ij}), \qquad i, j, k = 1, 2, \ldots, 9.$$

The bilinear operator in (23.43) has been written in symmetric form. It follows from (23.35) that

$$\|P''(x)\| \le \varpi_0 \max_{(i)} \sum_{j=1}^{9} b_{ij} = \varpi_0 \sum_{j=1}^{9} b_{9j}$$

$$= 0.69004017\varpi_0. \tag{23.44}$$

For $\varpi_0 = 0.1$, $x_0 = (1, 1, \ldots, 1)$, the following values were obtained by direct computation [13]:

$$\eta_0 = \|x_1 - x_0\| = 3.66110417 \times 10^{-2}, \tag{23.45}$$

$$B_0 = \|[P'(x_0)]^{-1}\| = 1.07322208, \tag{23.46}$$

and so from (23.44),

$$h_0 = 2.71129055 \times 10^{-3} \le \tfrac{1}{2}. \tag{23.47}$$

Thus the hypotheses of Theorem 22.4 are satisfied in this case. The fact that $h_0 \le \tfrac{1}{2}$ also allows us to obtain the error estimate

$$\|x^* - x_1\| \le 1.98526343 \times 10^{-4}. \tag{23.48}$$

The numerical solution of the systems (23.42) was carried out by Newton's method in the same fashion that successive substitutions were used in Section 13; namely, the solution for each value of ϖ_0, starting with $\varpi_0 = 0.1$ and $x_0 = (1, 1, \ldots, 1)$, is used as the initial approximation for the next larger value of ϖ_0. The iteration number N such that

$$\|x^* - x_N\| < 10^{-9} \tag{23.49}$$

is shown in Table 23.1 [13] for each value of ϖ_0.

Comparison of Table 23.1 with Table 13.4 gives a dramatic illustration of the power of Newton's method, especially for the larger values of ϖ_0. The results obtained by Newton's method agree with the values given in Table 13.3 to within one or two units in the last decimal place, except for $\varpi_0 = 1.0$. The values of ξ_i^N, $i = 1, 2, \ldots, 9$, found by Newton's method are given in Table 23.2, together with the differences

$$\Delta_i = \xi_i^N - \xi_i^S, \qquad i = 1, 2, \ldots, 9, \tag{23.50}$$

TABLE 23.1

**Number of iterations required to
obtain the solutions in Table 13.3
by Newton's method**

ϖ_0	N
0.1	2
0.2	2
0.3	3
0.4	3
0.5	3
0.6	3
0.7	3
0.8	3
0.9	4
1.0	13

where the values in Table 13.3, found by the method of successive substitutions, are denoted by ξ_i^S, $i = 1, 2, \ldots, 9$.

TABLE 23.2

**The values ξ_i^N found by Newton's
method for $\varpi_0 = 1.0$ and the
differences Δ_i, $i = 1, 2, \ldots, 9$**

i	ξ_i^N	Δ_i
1	1.05078690	0.00000013
2	1.20849425	0.00000075
3	1.43699363	0.00000211
4	1.71327544	0.00000440
5	2.01197592	0.00000765
6	2.30479316	0.00001161
7	2.56296985	0.00001573
8	2.76053567	0.00001927
9	2.87737350	0.00002153

Table 23.2 illustrates the danger of the use of the convergence of a numerical
process as a criterion of error. A more simple illustration is that the iterative
process

$$x_{n+1} = 0.999x_n, \qquad x_0 = 0.400, \tag{23.51}$$

gives, to three decimal places,

$$x_1 = 0.400, \tag{23.52}$$

and thus converges to this accuracy, even though

$$\|x^* - x_1\| = 0.400, \tag{23.53}$$

where $x^* = 0.000$ is the solution of the fixed-point problem

$$x = 0.999x. \tag{23.54}$$

EXERCISES 23

1. Express the total number of distinct coefficients

$$P''_{ijk} = \frac{\partial^2 P_i(\xi_1, \xi_2, \ldots, \xi_n)}{\partial \xi_j \, \partial \xi_k}, \qquad i, j, k = 1, 2, \ldots, n,$$

   of the second derivative $P''(x)$ of an arbitrary twice differentiable operator $P(x)$ in R^n.
2. Prove that (23.35) holds in R_∞^n. *Hint.* First extend the result of Problem 5, Exercises 8, to R_∞^n.
3. Obtain estimates (upper bounds) for $\|P''(x)\|$ in R_1^n and in R_p^n, $1 < p < \infty$.
4. Obtain an initial approximation x_0 to the solution x^* in the positive octant ($\xi > 0$, $\eta > 0$, $\zeta > 0$) of the equation $P(x) = 0$, where P is defined by (23.20), for which the hypotheses of Theorem 22.4 hold.
5. For the systems (23.42) suppose that the evaluation of $P(x)$ [or $F(x) = y + Bxx$] requires 25 msec and the calculation of $[P'(x)]^{-1}$ requires 150 msec. For what values of ϖ_0 is Newton's method more efficient in terms of total solution time than successive substitutions to calculate the values given in Table 13.3?

24. PROGRAMMING NEWTON'S METHOD FOR AUTOMATIC COMPUTATION

In common with the method of successive substitutions, we are generally required to evaluate the nonlinear operator $P(x)$ at each step of Newton's method. For the case of finite systems this amounts to the evaluation of the n functions

$$P_i(x) = P_i(\xi_1, \xi_2, \ldots, \xi_n), \qquad i = 1, 2, \ldots, n. \tag{24.1}$$

Under ordinary circumstances it is a fairly straightforward task to program an electronic digital computer to perform the indicated evaluations, at least to an acceptable degree of accuracy.

In addition to the above, the user of Newton's method must obtain the n^2 partial derivatives

$$\frac{\partial P_i(x)}{\partial \xi_j} = \frac{\partial P_i(\xi_1, \xi_2, \ldots, \xi_n)}{\partial \xi_j}, \qquad i, j = 1, 2, \ldots, n, \tag{24.2}$$

which are the elements of the Jacobian matrix $P'(x)$, and evaluate them at each step of the iteration. This evaluation is followed by a matrix inversion or the solution of a system of n linear algebraic equations to obtain the next approximation to the solution. Furthermore, if we wish to carry out convergence and error analysis on the basis of Theorem 22.4, a suitable bound is required for the norm of the second derivative $P''(x)$, which has as its elements the second partial derivatives

$$\frac{\partial^2 P_i(x)}{\partial \xi_j \, \partial \xi_k} = \frac{\partial^2 P_i(\xi_1, \xi_2, \ldots, \xi_n)}{\partial \xi_j \, \partial \xi_k}, \qquad i, j, k = 1, 2, \ldots, n. \qquad (24.3)$$

The determination of an upper bound of the form (23.35) for $\|P''(x)\|$ could be a substantial computational project in its own right.

Although the extra labor involved in the use of Newton's method appears to be formidable at first sight, it turns out that the entire process can be automated for functions (24.1) of the type ordinarily suitable for programming for machine evaluation. This includes obtaining the required derivatives and bounds, so that the user needs to supply coding only for the original functions (24.1) and the necessary control parameters. The purpose of this section is to describe the theory of an automatic computer program of this type [14]. Because of the wide variability in computing languages and machines, the discussion will not be given with a specific system in mind, but in general terms to permit adaptation to any particular situation.

To describe computer programming, it is convenient to introduce the *assignment operator* $:=$. The statement

$$x := y \qquad (24.4)$$

means that the value of y before the operation is performed is assigned to x as a result of the operation. As another example,

$$x := x + 1 \qquad (24.5)$$

means that the new value of x is the sum of the old value of x and 1. In executing the operation (24.5), the computer works much like a bookkeeper seated at a pigeonhole desk: A slip of paper with a number on it is taken from the slot labeled x, one is added to it, and the new total returned to the same location. Some computer languages use the equality sign $=$ to denote the assignment operator [15]; in such a case, the convention is usually adopted that the right-hand side of a statement refers to previous values, the assignment of the indicated result being made to the variable on the left-hand side of the statement.

Attention will now be turned to the problem of obtaining the derivatives (24.2) and (24.3) from the functions (24.1). Although programs for differentiation and other symbolic operations have become fairly widespread, it is worthwhile from a conceptual standpoint to describe the operation of a typical program of this type [16] written by Allen Reiter and developed further by Julia H. Gray, based on ideas proposed by R. E. Moore [17]. Other techniques are also feasible, but the discussion here will be limited to one approach which has proved to be successful.

The basic idea involved in the automatic differentiation program considered is to perform the evaluation of the given function by a "routine" (sequence of instructions) in which each instruction involves only one differentiable operation; for example, consider the function

$$f(x, y) = (xy + \sin x + 4)(3y^2 + 6). \qquad (24.6)$$

A typical statement for the evaluation of (24.6) would be

$$F := (X*Y + SINF(X) + 4.0)*(3.0*Y**2 + 6.0) \qquad (24.7)$$

In (24.7), $*$ denotes multiplication and $**$ exponentiation. The computer program would then analyze (24.7) and produce the following sequence of instructions:

$$
\begin{aligned}
T1 &:= X*Y, \\
T2 &:= SINF(X), \\
T3 &:= T1 + T2, \\
T4 &:= T3 + 4.0, \\
T5 &:= Y**2, \\
T6 &:= 3.0*T5, \\
T7 &:= T6 + 6.0, \\
F &:= T4*T7.
\end{aligned}
\qquad (24.8)
$$

Given the values of X and Y, the execution of the sequence of instructions (24.8) from top to bottom will result in the correct value being assigned to F. Basic functions such as SINF(X) are normally evaluated by standard subroutines. The important point is that the rules for differentiating all of the operations involved are known and can be stored in the computer, so that term-by-term differentiation of (24.8) can be accomplished; for example, the "dictionary" entry for the differentiation of an instruction of the form

$$U3 := U1*U2 \qquad (24.9)$$

would be

$$
\begin{aligned}
V1 &:= U1*DU2, \\
V2 &:= DU\,1*U2, \\
DU\,3 &:= V1 + V2.
\end{aligned}
\qquad (24.10)
$$

· Similarly, for

$$U2 := SINF(U1),$$

(24.11)

we would have

$$V1 := COSF(U1),$$
$$DU\,2 := V1*DU\,1.$$

(24.12)

Applying this technique to (24.8), we would obtain

$$V1 := X*DY,$$
$$V2 := DX*Y,$$
$$DT1 := V1 + V2,$$
$$V3 := COSF(X),$$
$$DT2 := V3*DX,$$
$$DT3 := DT1 + DT2,$$
$$DT4 := DT3,$$
$$V4 := Y**1,$$
$$V5 := 2.0*V4,$$
$$DT5 := V5*DY,$$
$$DT6 := 3.0*DT5,$$
$$DT7 := DT6,$$
$$V6 := T4*DT7,$$
$$V7 := DT4*T7,$$
$$DF := V6 + V7.$$

(24.13)

It is assumed that the program recognizes constants as such. To obtain the partial derivative DXF of F with respect to X, the program sets DX = 1, DY = 0 in (24.13), and after reduction of trivialities finds

$$DXT1 := Y,$$
$$DXT2 := COSF(X),$$
$$DXT4 := DXT1 + DXT2,$$
$$DXF := DXT4*T7.$$

(24.14)

Comparison with (24.6) and (24.8) shows that (24.14) will evaluate the correct result

$$\frac{\partial f}{\partial x} = (y + \cos x)(3y^2 + 6),$$

(24.15)

assuming that F has been evaluated previously. Similarly, the partial derivative of F with respect to Y may be found from (24.13) by setting $DX = 0$, $DY = 1$. Since (24.14) is a *code list* of the same form as (24.13), it can be differentiated in turn to yield higher partial derivatives.

As an example of the actual application of the program CODEX [16] (used in this case as a subprogram of NEWTON [14]), the functions

$$F_1(x, y, z) = 16x^4 + 16y^4 + z^4 - 16,$$
$$F_2(x, y, z) = x^2 + y^2 + z^2 - 3, \qquad (24.16)$$
$$F_3(x, y, z) = x^3 - y,$$

were coded for the machine [13] as

$$F1 = 16.*X**4 + 16.*Y**4 + Z**4 - 16. \quad \$$$
$$F2 = X**2 + Y**2 + Z**2 - 3. \quad \$ \qquad (24.17)$$
$$F3 = X**3 - Y \quad \$$$

The character set for this machine uses $=$ for $:=$, and the $\$$ sign indicates the end of an equation. For

$$\xi_1 = x, \qquad \xi_2 = y, \qquad \xi_3 = z, \qquad (24.18)$$

the notation $i, j, 0$ in the following output stands for $\partial F_i / \partial \xi_j$, $i, j = 1, 2, 3$, while i, j, k signifies $\partial^2 F_i / \partial \xi_j \, \partial \xi_k$, $i, j, k = 1, 2, 3$. In addition to preparing the code lists, the program CODEX [16] generates internal machine language routines for the actual evaluation of the functions and derivatives. Programs that translate symbolic statements into actual machine instructions are called *compilers*. The output obtained gives the code lists for the functions and the partial derivatives; the message NOT ON F-LIST means that the given function (or derivative) does not contain the indicated variable, so that the corresponding partial derivative is zero. Note that the program obtains only the distinct second derivatives.

F1 = 16.*X**4 + 16.*Y**4 + Z**4 - 16. $

CODE LIST FOR F_1

```
          0001T = X        **   0004C
          0002T = 0012C    *    0001T
          0003T = Y        **   0004C
          0004T = 0012C    *    0003T
          0005T = 0002T    +    0004T
          0006T = Z        **   0004C
          0007T = 0005T    +    0006T
      F1      = 0007T      -    0012C
```

F2 = X**2 + Y**2 + Z**2 - 3. $

CODE LIST FOR F₂

$$
\begin{aligned}
0010T &= X   &&** \quad 0002C \\
0011T &= Y   &&** \quad 0002C \\
0012T &= 0010T &&+ \quad 0011T \\
0013T &= Z   &&** \quad 0002C \\
0014T &= 0012T &&+ \quad 0013T \\
F2 &= 0014T   &&- \quad 0003C
\end{aligned}
$$

F3 = X**3 − Y $

CODE LIST FOR F₃

$$
\begin{aligned}
0015T &= X   &&** \quad 0003C \\
F3 &= 0015T   &&- \quad Y
\end{aligned}
$$

CODE LIST FOR 1, 1, 0

$$
\begin{aligned}
0016T &= X   &&** \quad 0003C \\
0017T &= 0004C &&* \quad 0016T \\
0020T &= 0012C &&* \quad 0017T \\
011 &=   &&+ \quad 0020T
\end{aligned}
$$

CODE LIST FOR 1, 2, 0

$$
\begin{aligned}
0021T &= Y   &&** \quad 0003C \\
0022T &= 0004C &&* \quad 0021T \\
0023T &= 0012C &&* \quad 0022T \\
012 &=   &&+ \quad 0023T
\end{aligned}
$$

CODE LIST FOR 1, 3, 0

$$
\begin{aligned}
0024T &= Z   &&** \quad 0003C \\
0025T &= 0004C &&* \quad 0024T \\
013 &=   &&+ \quad 0025T
\end{aligned}
$$

CODE LIST FOR 1, 1, 1

$$
\begin{aligned}
5671T &= X   &&** \quad 0002C \\
5672T &= 0003C &&* \quad 5671T \\
5673T &= 0004C &&* \quad 5672T \\
5674T &= 0012C &&* \quad 5673T \\
111 &=   &&+ \quad 5674T
\end{aligned}
$$

CODE LIST FOR 1, 1, 2
112 NOT ON F-LIST.

CODE LIST FOR 1, 1, 3
113 NOT ON F-LIST.

CODE LIST FOR 1, 2, 2

$$
\begin{aligned}
5671T &= Y   &&** \quad 0002C \\
5672T &= 0003C &&* \quad 5671T \\
5673T &= 0004C &&* \quad 5672T \\
5674T &= 0012C &&* \quad 5673T \\
122 &=   &&+ \quad 5674T
\end{aligned}
$$

CODE LIST FOR 1, 2, 3
123 NOT ON F-LIST.

CODE LIST FOR 1, 3, 3
```
              5671T = Z        **   0002C
              5672T = 0003C     *   5671T
              5673T = 0004C     *   5672T
                133 =           +   5673T
```

CODE LIST FOR 2, 1, 0
```
              0026T = 0002C     *   X
                021 =           +   0026T
```

CODE LIST FOR 2, 2, 0
```
              0027T = 0002C     *   Y
                022 =           +   0027T
```

CODE LIST FOR 2, 3, 0
```
              0030T = 0002C     *   Z
                023 =           +   0030T
```

CODE LIST FOR 2, 1, 1
```
                211 =           +   0002C
```

CODE LIST FOR 2, 1, 2
```
              212 NOT ON F-LIST.
```

CODE LIST FOR 2, 1, 3
```
              213 NOT ON F-LIST.
```

CODE LIST FOR 2, 2, 2
```
                222 =           +   0002C
```

CODE LIST FOR 2, 2, 3
```
              223 NOT ON F-LIST.
```

CODE LIST FOR 2, 3, 3
```
                233 =           +   0002C
```

CODE LIST FOR 3, 1, 0
```
              0031T = X         **   0002C
              0032T = 0003C      *   0031T
                031 =            +   0032T
```

CODE LIST FOR 3, 2, 0
```
                032 =           −   0001C
```

CODE LIST FOR 3, 3, 0
```
              033 NOT ON F-LIST.
```

CODE LIST FOR 3, 1, 1
```
              5671T = 0002C      *   X
              5672T = 0003C      *   5671T
                311 =            +   5672T
```

CODE LIST FOR 3, 1, 2
```
              312 NOT ON F-LIST.
```

CODE LIST FOR 3, 1, 3
```
              313 NOT ON F-LIST.
```

CODE LIST FOR 3, 2, 2
        322 NOT ON F-LIST.

CODE LIST FOR 3, 2, 3
        323 NOT ON F-LIST.

CODE LIST FOR 3, 3, 3
        333 NOT ON F-LIST.

Once the partial derivatives and the programs for evaluating them have been produced by the computer, the evaluation of $P'(x)$ and the generation of the Newton sequence become routine computational procedures. The problems of convergence analysis and error estimation remain. In R^n_∞ the calculation of the quantity

$$\eta_0 = \|x_1 - x_0\| = \max_{(i)} |\xi_i^{(1)} - \xi_i^{(0)}| \tag{24.19}$$

and for

$$G_0 = (g_{ij}^{(0)}) = [P'(x_0)]^{-1} \tag{24.20}$$

the determination of

$$B_0 = \|G_0\| = \max_{(i)} \sum_{j=1}^{n} |g_{ij}^{(0)}| \tag{24.21}$$

are perfectly straightforward. To simplify obtaining a satisfactory upper bound for $\|P''(x)\|$ in the general case it is expedient to use the following theorem.

Theorem 24.1. If

$$\|P''(x)\| \le K \tag{24.22}$$

for $x \in U(x_0, 2\eta_0)$ and

$$h_0 = B_0 \eta_0 K \le \tfrac{1}{2}, \tag{24.23}$$

then the equation $P(x) = 0$ has a unique solution $x = x^*$ in $U(x_0, 2\eta_0)$ to which the Newton sequence converges; furthermore,

$$\|x^* - x_1\| \le 2h_0 \eta_0. \tag{24.24}$$

PROOF. The existence of x^* and the convergence of the Newton sequence to it follow directly from Theorem 22.4 since, if $h_0 \le \tfrac{1}{2}$, then

$$2\eta_0 \ge r_0 = \frac{1 - \sqrt{1 - 2h_0}}{h_0} \eta_0, \tag{24.25}$$

so that (22.18) is satisfied. The uniqueness of x^* in $U(x_0, 2\eta_0)$ follows immediately from Theorems 22.6 and 22.7. The error bound (24.24) is a consequence of Theorem 22.5.

Since x_1 will be available along with η_0 and B_0 in the computational case, the use of the error bound (24.24) is more efficient than employing the estimate

$$\|x^* - x_0\| \leq 2\eta_0 \tag{24.26}$$

for the error of x_0 as an approximation to x^*.

The practical virtue of Theorem 24.1 is that the problem of finding an upper bound for $\|P''(x)\|$ is set in the specific ball

$$\bar{V}_0 = \bar{U}(x_0, 2\eta_0). \tag{24.27}$$

Assuming that P'' is continuous on $\bar{V}_0$, (23.35) becomes

$$\|P''(x)\| \leq \max_{x \in \bar{V}_0} \left[\max_{(i)} \sum_{\substack{j=1 \\ k=1}}^{n} |P''_{ijk}(x)| \right]. \tag{24.28}$$

It follows at once from (24.28) that

$$\|P''(x)\| \leq \max_{(i)} \sum_{\substack{j=1 \\ k=1}}^{n} \max_{x \in \bar{V}_0} |P''_{ijk}(x)|. \tag{24.29}$$

The formulation (24.29) reduces the problem of finding an upper bound for $\|P''(x)\|$ to finding estimates M_{ijk} such that

$$M_{ijk} \geq \max_{x \in \bar{V}_0} |P''_{ijk}(x)| \tag{24.30}$$

for the distinct individual second partial derivatives found by the computer differentiation program. This can be done automatically for the type of operator considered by the use of *interval arithmetic* [17,18].

In essence, interval arithmetic uses the bounded closed intervals of real numbers

$$\tilde{x} = [x_L, x_R], \qquad x_L \leq x_R, \tag{24.31}$$

as elements. Interval arithmetic operations are defined by

$$\tilde{x} \, o \, \tilde{y} = \{z : z = x \, o \, y, x \in \tilde{x}, y \in \tilde{y}\}, \tag{24.32}$$

where o denotes addition, subtraction, multiplication, or division. Division by an interval containing zero is excluded. If $g(x)$ is a continuous real function of x, then it is possible to define the corresponding *interval function* $\tilde{g}$ by

$$\tilde{g}(\tilde{x}) = \{y : y = g(x), x \in \tilde{x}\}, \tag{24.33}$$

provided the result is also a finite interval. Thus the operations of interval arithmetic can be extended to include functions such as $\sin x$, e^x, etc. The basic operations of interval arithmetic, together with a selected set of interval functions, will be called an *extended interval arithmetic*. A real function $g(x)$ that can be evaluated by a routine consisting of a code list of the form (24.8) which uses only operations in the extended interval arithmetic available is called an *interval computable function*. For a given interval $\tilde{x}$, if the instructions in a code list for evaluating the function $g(x)$ are executed in extended interval arithmetic, then the result will be an interval called the *interval evaluation* of $g(x)$, which will be denoted by $[g(\tilde{x})]$, provided that it exists. The interval evaluation of a real function differs, in general, from the corresponding interval function. For example, if

$$g(x) = x - x^2 \tag{24.34}$$

and $\tilde{x} = [0, 1]$,

$$\tilde{g}([0, 1]) = [0, \tfrac{1}{4}] \tag{24.35}$$

for the corresponding interval function by (24.33). A typical code list to evaluate $g(x)$ would be:

$$\begin{aligned} \text{T1} &:= \text{X} * \text{X}, \\ \text{G} &:= \text{T1} - \text{X}. \end{aligned} \tag{24.36}$$

Applying interval arithmetic to (24.36) gives

$$[g([0, 1])] = [-1, 1]. \tag{24.37}$$

The interval evaluation of a function will generally depend on the operations available in the extended interval arithmetic available as well as on the particular code list used for evaluation of the function. However, (24.35) and (24.37) suggest the general result due to R. E. Moore [18].

Theorem 24.2 (*The Fundamental Theorem of Interval Analysis*).   If $g(x)$ is an interval computable function, then

$$\tilde{g}(\tilde{x}) \subset [g(\tilde{x})] \tag{24.38}$$

for any interval $\tilde{x}$ for which both $\tilde{g}(\tilde{x})$ and $[g(\tilde{x})]$ exist.

This result extends immediately to real functions $g(x_1, x_2, \ldots, x_n)$ of several variables. An interval vector $\tilde{x}$ is defined by

$$\tilde{x} = (\tilde{x}_1, \tilde{x}_2, \ldots, \tilde{x}_n) = \{x : x = (x_1, x_2, \ldots, x_n), x_i \in \tilde{x}_i, i = 1, 2, \ldots, n\}. \tag{24.39}$$

The corresponding interval function $\tilde{g}(\tilde{x})$ to a real vector function $g(x) = g(x_1, x_2, \ldots, x_n)$ and the interval evaluation $[g(\tilde{x})]$ of $g(x)$ are defined by strict analogy to the single variable case, and Theorem 24.2 holds also for interval computable vector functions [18].

For the problem of obtaining the upper bound (24.29), the ball $\bar{V}_0 = \bar{U}(x_0, 2\eta_0)$ in R_∞^n corresponds to the interval vector $\tilde{x} = (\tilde{x}_1, \tilde{x}_2, \ldots, \tilde{x}_n)$ defined by

$$\tilde{x}_i = [\xi_i^{(0)} - 2\eta_0, \xi_i^{(0)} + 2\eta_0], \qquad i = 1, 2, \ldots, n. \tag{24.40}$$

Assuming that the code lists for the second derivatives have been computed by the differentiation subprogram as in the foregoing example, a subprogram for interval arithmetic can be employed to obtain the interval evaluations of the second derivatives

$$[P''_{ijk}(\tilde{x})] = \left[\frac{\partial^2 P_i(\tilde{x})}{\partial \xi_j \, \partial \xi_k}\right], \qquad i, j, k = 1, 2, \ldots, n. \tag{24.41}$$

For this purpose the automatic program NEWTON [14] uses as a subprogram the interval arithmetic program INTERVAL [19], which was written by Allen Reiter. Writing

$$[P''_{ijk}(\tilde{x})] = [a_{ijk}, b_{ijk}], \tag{24.42}$$

we have by Theorem 24.2 as extended to vectors that

$$a_{ijk} \leq P''_{ijk}(x) \leq b_{ijk} \tag{24.43}$$

for $x \in \bar{V}_0$, $i, j, k = 1, 2, \ldots, n$. Consequently,

$$|P''_{ijk}(x)| \leq \max \{|a_{ijk}|, |b_{ijk}|\} = M_{ijk} \tag{24.44}$$

for $x \in \bar{V}_0$ and, thus, by (24.29) and (24.30), if

$$K = \max_{(i)} \sum_{\substack{j=1 \\ k=1}}^{n} M_{ijk}, \tag{24.45}$$

then

$$\|P''(x)\| \leq K \tag{24.46}$$

for $x \in \bar{V}_0$. The value (24.45) for K is somewhat crude, as it is usually a gross overestimate for the actual upper bound of $\|P''(x)\|$ on $\bar{V}_0$. It has the virtue, however, of being a rigorous upper bound for the second derivative which is obtainable by automatic computation. After K has been calculated, all that remains to be done to verify the hypotheses of Theorem 24.1 is to compute h_0 as defined by (24.33). Once a satisfactory value of h_0 is obtained, that is, $h_0 \leq \frac{1}{2}$, it is no longer necessary to recompute K at each step, so that some labor may be avoided after convergence is assured. The current values of η_0 and B_0 can be used to calculate h_0 for the error bound (24.24) at each step of the iteration.

The following output shows the application of the automatic program NEWTON [14] to the solution of the system

$$F_i(x, y, z) = 0, \qquad i = 1, 2, 3, \tag{24.47}$$

where the functions F_i are those defined by (24.16). This system represents the intersection of a hyperellipsoid, a sphere, and a cylinder. By sketching them, the point $x_0 = (1, 1, 1)$ is taken as the initial approximation to the solution of this system in the positive octant. The program began the iteration process after producing the necessary derivatives given in the foregoing.

NEWTONS METHOD
X 1. Y 1. Z 1. $

(The program listed the functions and derivatives at this point.)

ITERATION NUMBER   0

NORM F =   1.70000000+001

X   =   1.00000000+000   F1 =   1.70000000+001
Y   =   1.00000000+000   F2 =   0.00000000+000
Z   =   1.00000000+000   F3 =   0.00000000+000

TIME PER ITERATION = 120.00 MILLISECONDS

ITERATION NUMBER   1

NORM F =   4.79191714+000
NORM CX =   2.83333333−001
BOUND F PRIME INVERSE =   5.50000000−001
BOUND F DBL PRIME =   9.71960000+002
  HO = 1.51463767+002
X   =   9.29166667−001   F1 =   4.79191714+000
Y   =   7.87500000−001   F2 =   1.30451389−001
Z   =   1.28333333+000   F3 =   1.46966869−002

TIME PER ITERATION = 827.00 MILLISECONDS

ITERATION NUMBER   2

NORM F =   6.45309522−001
NORM CX =   9.43241407−002
BOUND F PRIME INVERSE =   5.56439198−001
BOUND F DBL PRIME =   4.48856898+002
  HO = 2.35585457+001
X   =   8.87074529−001   F1 =   6.45309522−001
Y   =   6.93175859−001   F2 =   1.20773905−002
Z   =   1.32086464+000   F3 =   4.86417094−003

TIME PER ITERATION = 797.00 MILLISECONDS

ITERATION NUMBER   3

NORM F =   1.84509452−002
NORM CX =   1.59811520−002
BOUND F PRIME INVERSE =   6.34254336−001
BOUND F DBL PRIME =   2.85088868+002
  HO = 2.88969353+000
X   =   8.78244398−001   F1 =   1.84509452−002
Y   =   6.77194707−001   F2 =   4.28336556−004
Z   =   1.33060980+000   F3 =   2.06810364−004

TIME PER ITERATION = 2181.00 MILLISECONDS

ITERATION NUMBER   4

NORM F =   1.47977844−005
NORM CX =   4.37403606−004
BOUND F PRIME INVERSE =   6.61437802−001
BOUND F DBL PRIME =   2.57938953+002
  HO =   7.46256805−002
ERROR =   6.52830835−005
X   =   8.77965993−001   F1   =   1.47977844−005
Y   =   6.76757304−001   F2   =   3.29047907−007
Z   =   1.33085521+000   F3   =   2.04207026−007

TIME PER ITERATION = 893.00 MILLISECONDS

SUCCESSFUL CONVERGENCE AT ITERATION NUMBER   5

WITH ERROR =   3.79255563−011 LESS THAN 5.00000000−009
X   =   8.77965760−001   F1   =   4.65661287−010
Y   =   6.76756971−001   F2   =   5.82076609−011
Z   =   1.33085541+000   F3   =   0.00000000+000

TOTAL ITERATION TIME   5959

In this listing the numerical notation $1.70000000 + 001$ means $1.7 \times 10^{+1} = 17$, other floating-point numbers being expressed in a similar fashion. The quantities necessary for the verification of the hypotheses of Theorem 24.1 are

$$\eta_0 = \text{NORM CX,}$$
$$B_0 = \text{BOUND F PRIME INVERSE,} \qquad (24.48)$$
$$K = \text{BOUND F DBL PRIME.}$$

Note that the convergence conditions are not satisfied until (x_3, y_3, z_3) has been used as the initial approximation, and that two more iterations give the result (x_5, y_5, z_5) which is accurate to eight decimal places on the basis of the error bound (24.24). This example shows that it is possible to do a certain amount of computation on faith in the eventual convergence of the process. The prudent programmer, however, will set limits on the total number of iterations and the amount of computer time used, as well as on the size of certain quantities, so that the program will not flounder aimlessly for an indefinite period if a solution is not accessible from the chosen initial point. In addition, *numerical convergence* can occur when $\|P(x)\|$ or $\|x_1 - x_0\|$ become extremely small, say of the order of magnitude of the roundoff error of the arithmetic operations in the machine being used. At this point further computation is usually a waste of time, and the uncertainty in η_0 may make the error bound (24.24) unreliable. The smallness of $\|P(x)\|$ or of $\|x_1 - x_0\|$ may also be used as a criterion for ending the iteration in place of the time-consuming error analysis previously described. This may often be satisfactory in practice, especially if the error estimation has been done for a typical case of the type of systems being solved. As discussed at the end of the previous section, however, sometimes a misleading impression of accuracy results.

Figures 24.1 through 24.3 give a geometric representation, called a *flow chart*, of the automatic program described in this section. The various significant parameters used for the control of the program are indicated on the flow chart. The actual values of these parameters for this example were

$$
\begin{aligned}
F &= \ 5.00000000-009 \\
FF &= \ 1.00000000+006 \\
BB &= \ 1.00000000+006 \\
C &= \ 5.00000000-009 \\
CC &= \ 1.00000000+006 \\
TLIM &= 1.20000000+005 \\
\end{aligned}
$$

IERR = 1 IMAT = 1 IDMP = 1
NITDIS = 1 LIMIT = 25 IAGAIN = 1
NOF = 1 NPRT = 0 IAVL = 1

$$
\begin{aligned}
E &= \ 5.00000000-009 \\
HH &= \ 1.00000000+006 \\
BNORM &= \ -0.00000000+000 \\
\end{aligned}
$$

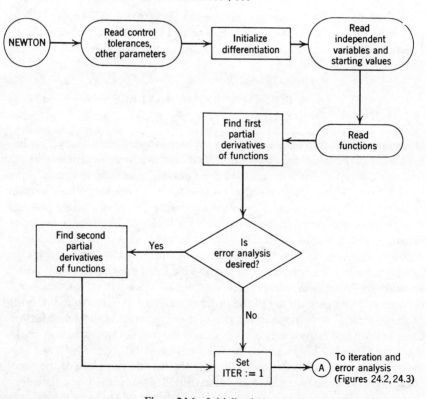

Figure 24.1 Initialization.

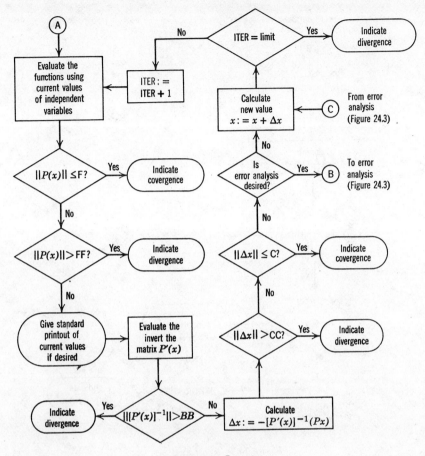

Figure 24.2 Iteration.

Some of the foregoing parameters control printing and other program functions and are not shown on the flow chart. For a complete description of the output listings given, the report [14] should be consulted.

The assumption on which the error analysis portion of the particular program described is based is that the quantities B_0, η_0, K are obtained exactly, or are at least upper bounds for the true values which would be obtained if all calculations were performed without error. In practice, especially if automatic calculation by interval arithmetic is used, the value of K is sufficiently overestimated so that the calculated value of h_0 will exceed the theoretical value in most cases, even if B_0, η_0 are underestimated due to computational errors. Thus we may usually rely on the error bound (24.24), at least outside of the region of magnitude of the computational error in the

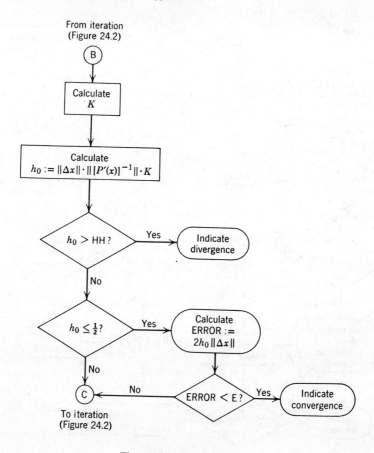

Figure 24.3 Error analysis.

value found for x_1. A more rigorous discussion of error estimation in the case that all operations are assumed to be inexact will be given in a later section.

EXERCISES 24

1. Write the code list for a "dictionary" differentiation of

$$U3 := U1**U2$$

(that is, $u_3 = u_1{}^{u_2}$), assuming that a standard subroutine LNF(U) is available for the evaluation of the natural logarithm ln u.

2. If F is a differentiable operator and $x_1, x_2 \in \bar{U}(x_0, r)$, then

$$\|F(x_1) - F(x_2)\| \leq \left\{ \sup_{x \in \bar{U}(x_0, r)} \|P'(x)\| \right\} \|x_1 - x_2\|.$$

On the basis of this result, discuss an automatic program for solving

$$x = F(x)$$

in R^n_∞ by successive substitutions, assuming that programs for differentiation and interval arithmetic are available.

3. Define the interval function $\tilde{g}(\tilde{x})$ corresponding to

$$g(x) = x - x^2.$$

4. Define the interval functions corresponding to e^x, $\sin x$, $\ln x$.

25. APPLICATION OF NEWTON'S METHOD TO INTEGRAL AND DIFFERENTIAL EQUATIONS

Effective methods for the solution of integral and differential equations are of great practical importance and there is a vast literature devoted to this subject. In this section the application of Newton's method to this problem is illustrated by means of a few typical examples.

A basic problem for the ordinary differential equation

$$\frac{dy}{dt} - f(t, y) = 0 \tag{25.1}$$

is to find a solution $y = y(t)$ in $C'[0, T]$ such that

$$y(0) = c. \tag{25.2}$$

This is called the *initial value problem* for (25.1). Corresponding to (25.1) is the operator

$$P(y) = \frac{dy}{dt} - f(t, y) \tag{25.3}$$

from $C'[0, T]$ into $C[0, T]$, assuming that f is continuous for $0 \leq t \leq T$ and $y \in C'[0, T]$. If f is differentiable with respect to y, then at $y_0 = y_0(t)$, we have that the derivative of P is

$$P'(y_0) = \frac{d}{dt} - f_2'(t, y_0(t))I, \tag{25.4}$$

a linear differential operator from $C'[0, T]$ into $C[0, T]$, where

$$f_2'(t, y_0(t)) = \left. \frac{\partial f(t, y)}{\partial y} \right|_{y=y_0(t)}, \qquad 0 \leq t \leq T. \tag{25.5}$$

In order to generate the Newton sequence of successive approximations $\{y_m\}$ to a solution of (25.1), starting with $y_0 = y_0(t)$, we must be able to invert linear differential operators of the form (25.4) or, what is the same, solve the linear differential equations

$$P'(y_m)(y_{m+1} - y_m) = -P(y_m), \qquad m = 0, 1, \ldots, \qquad (25.6)$$

for $y_{m+1} = y_{m+1}(t)$, $0 \le t \le T$, subject to the initial condition

$$y_{m+1}(0) = c. \qquad (25.7)$$

If we set

$$u_m(t) = y_{m+1}(t) - y_m(t),$$

$$a_m(t) = -f_2'(t, y_m(t)), \qquad (25.8)$$

$$v_m(t) = -P(y_m) = -\frac{dy_m(t)}{dt} + f(t, y_m(t)),$$

then (25.6) becomes

$$\frac{du_m(t)}{dt} + a_m(t) u_m(t) = v_m(t), \qquad (25.9)$$

with the initial condition

$$u_m(0) = 0, \qquad m = 0, 1, \ldots. \qquad (25.10)$$

(It is assumed that $y_0(0) = c$.) The solution of this initial value problem is well known [20]. Define

$$A_m(t) = \int_0^t a_m(s)\, ds, \qquad 0 \le t \le T, \qquad (25.11)$$

and multiply both sides of (25.9) by

$$\exp\{A_m(t)\} = e^{A_m(t)}, \qquad (25.12)$$

which gives

$$\frac{d}{dt}(e^{A_m(t)} u_m(t)) = e^{A_m(t)} v_m(t), \qquad 0 \le t \le T. \qquad (25.13)$$

Integrating both sides of (25.13) yields

$$u_m(t) = \int_0^t e^{A_m(s) - A_m(t)} v_m(s)\, ds, \qquad 0 \le t \le T. \qquad (25.14)$$

Comparison of (25.14) with (25.9) shows that the inverse of the linear differential operator

$$L = \frac{d}{dt} + a_m(t)I \qquad (25.15)$$

from $C_0'[0, T]$ [the subspace of $C'[0, T]$ consisting of functions $u = u(t)$

satisfying (25.10)] into $C[0, T]$ is the linear integral operator K_m defined by

$$u = K_m v = \int_0^t K_m(t, s)\, v(s)\, ds, \qquad 0 \le t \le T, \quad v \in C[0, T], \quad (25.16)$$

with

$$K_m(t, s) = e^{A_m(s) - A_m(t)}, \qquad 0 \le s, t \le T. \qquad (25.17)$$

The formula for generating the Newton sequence of successive approximations $y_m(t)$ to the solution of (25.1) is thus

$$y_{m+1}(t) = y_m(t) + \int_0^t e^{A_m(s) - A_m(t)} v_m(s)\, ds, \qquad 0 \le t \le T, \quad m = 0, 1, 2, \ldots, \qquad (25.18)$$

where v_m, A_m are given by (25.8) and (25.11) respectively, and it is assumed that $y_0(0) = 0$.

As a specific example consider the initial value problem

$$\frac{dy}{dt} - (1 + y^2) = 0,$$

$$y(0) = 0. \qquad (25.19)$$

Here

$$a_m(t) = 2 y_m(t),$$

$$v_m(t) = \frac{dy_m}{dt} + (1 + y_m^2), \qquad (25.20)$$

so that for

$$y_0(t) = t \qquad (25.21)$$

we have

$$v_0(t) = t^2,$$

$$A_0(t) = t^2, \qquad (25.22)$$

and, thus, by (25.18),

$$y_1(t) = t + \int_0^t e^{s^2 - t^2} s^2\, ds. \qquad (25.23)$$

An explicit formula for the integral (25.24) cannot be given in terms of elementary functions; however, the integrand can be expanded in a power series which converges uniformly for all s and integrated term-by-term to obtain

$$y_1(t) = t + \frac{t^3}{3} + \frac{2t^5}{15} + \frac{4t^7}{105} + \frac{8t^9}{945} + \cdots. \qquad (25.24)$$

This result could be substituted into (25.18) to obtain an infinite series expansion for $y_2(t)$, but the calculations would be exceedingly complex.

If $T < \pi/2$, then the problem (25.19) has the solution

$$y^*(t) = \tan t, \qquad 0 \le t \le T. \tag{25.25}$$

Comparison of the series (25.24) with the Maclaurin series for $\tan t$ gives

$$y^*(t) - y_1(t) = \frac{t^7}{61} + \frac{38t^9}{2835} + \cdots, \qquad 0 \le t \le T < \frac{\pi}{2}. \tag{25.26}$$

Thus, $y_1(t)$ will be a good approximation to $y^*(t)$ for small t.

This simple example is instructive in two respects. First of all, it shows that an analytic attack on the problem quickly becomes cumbersome. Second, even though the linear differential equations (25.6) defining the Newton sequence have solutions in $C'[0, T]$ for all T, the original nonlinear differential equation (25.19) has the solution (25.25) in $C'[0, T]$ only if $T < \pi/2$.

Another approach to the solution of the initial value problem is the *integral equation technique*. Integral and differential equations are intimately related [21,22] and it is often possible to simplify the treatment of various aspects of a problem by expressing it in one or the other of these forms. By direct integration, (25.1) and (25.2) become the single equation

$$y(t) - c - \int_0^t f(s, y(s)) \, ds = 0, \tag{25.27}$$

a nonlinear integral equation of the *Volterra type*, so-called because the upper limit of integration is variable. If the operator $P(y)$ in $C[0, T]$ is defined by

$$P(y) = y(t) - c - \int_0^t f(s, y(s)) \, ds, \tag{25.28}$$

then its derivative $P'(y_0)$ at $y_0 = y_0(t)$ will be a linear Volterra integral operator of the form

$$P'(y_0) = I - V_0, \tag{25.29}$$

with the kernel

$$V_0(s, t) = f_2'(s, y_0(s))$$

[(see (10.14) through (10.17)]. Thus the function $u_m(t)$ defined by (25.8) satisfies the linear Volterra integral equation

$$u_m(t) - \int_0^t f_2'(s, y_m(s)) \, u_m(s) \, ds = w_m(t), \qquad 0 \le t \le T, \tag{25.30}$$

where

$$w_m(t) = -y_m(t) + c + \int_0^t f(s, y_m(s)) \, ds. \tag{25.31}$$

As indicated in Chapter 1, the linear Volterra integral operator

$$P'(y_m) = I - V_m \tag{25.32}$$

has the inverse

$$[P'(y_m)]^{-1} = \sum_{n=0}^{\infty} V_m{}^n \qquad (25.33)$$

[see (10.24)], and by (10.23),

$$\|[P'(y_m)]^{-1}\| \le e^{MT}, \qquad (25.34)$$

where M is any constant such that

$$|V_m(s, t)| \le M, \qquad 0 \le s, t \le T. \qquad (25.35)$$

For the simple example (25.19), (25.30) becomes, for $y_0(t) = t$,

$$u_0(t) - \int_0^t 2s \, u_0(s) \, ds = \frac{t^3}{3}. \qquad (25.36)$$

Solving (25.36) by iteration gives the power series

$$u_0(t) = \frac{t^3}{3} + \frac{2t^5}{15} + \frac{4t^7}{105} + \cdots, \qquad (25.37)$$

so that $y_1(t)$ obtained in this way agrees exactly with (25.24).

The second derivative of the differential operator (25.3) at $y = y_0(t)$ is

$$P''(y_0) = -f''_2(t, y_0(t))I_2, \qquad (25.38)$$

where I_2 is the bilinear operator corresponding to multiplication, that is, for $x, y \in C'[0, T]$,

$$I_2xy = x(t) \, y(t), \qquad 0 \le t \le T. \qquad (25.39)$$

It is assumed that $f(t, y)$ is twice differentiable with respect to y. (A Banach space such as $C'[0, T]$, in which the product xy of elements is defined and $\|xy\| \le \|x\| \cdot \|y\|$, is called a *Banach algebra*.)

For the integral operator (25.28) we have

$$P''(y_0) = -f''_2(s, y_0(s))I, \qquad (25.40)$$

where the product

$$P''(y_0)x = -f''_2(s, y_0(s)) \, x(s) \qquad (25.41)$$

is to be regarded as the kernel of the linear Volterra integral transformation

$$[P''(y_0)x]y = \int_0^t [-f''_2(s, y_0(s)) \, x(s)] \, y(s) \, ds, \qquad 0 \le t \le T. \quad (25.42)$$

Note that although (25.38) and (25.40) involve exactly the same function

$$f''_2(t, y_0(t)) = \frac{\partial^2 f(t, y)}{\partial y^2}\bigg|_{y=y_0(t)}, \qquad 0 \le t \le T, \qquad (25.43)$$

they are conceptually different operators.

For computational purposes, Noble [23] has pointed out that it is better to truncate the series (25.24) for $y_1(t)$ at what may be considered to be its final significant term to obtain a new initial approximation, rather than trying to find $y_2(t)$ by manipulation of the entire series for $y_1(t)$. Thus since t seems to be an accurate first term, we could take

$$y_0(t) = t + \frac{t^3}{3} \qquad (25.44)$$

on the assumption that the number of significant terms doubles at each step of Newton's method. Actually, (25.26) shows that it would be safe to take

$$y_0(t) = t + \frac{t^3}{3} + \frac{2t^5}{15}. \qquad (25.45)$$

Another important class of problems for ordinary differential equations consists of what are called *boundary value problems*. A typical example is to find a solution $y \in C''[0, 1]$ of the second-order differential equation

$$\frac{d^2y}{dx^2} - f(x, y) = 0 \qquad (25.46)$$

which satisfies the *boundary conditions*

$$y(0) = y(1) = 0. \qquad (25.47)$$

It will be assumed that $f(x, y)$ is continuous in x, $0 \leq x \leq 1$, and twice continuously differentiable with respect to y, so that

$$P(y) = \frac{d^2y}{dx^2} - f(x, y) \qquad (25.48)$$

is a twice differentiable operator from $C''[0, 1]$ into $C[0, 1]$. The first two derivatives of P at $y = y_m(x)$ are

$$P'(y_m) = \frac{d^2}{dx^2} - f'_2(x, y_m(x))I \qquad (25.49)$$

and

$$P''(y_m) = -f''_2(x, y_m(x))I_2 \qquad (25.50)$$

respectively. Setting

$$u_m(x) = y_{m+1}(x) - y_m(x),$$

$$v_m(x) = -P(y_m) = -\frac{d^2y_m(x)}{dx^2} + f(x, y_m(x)), \qquad m = 0, 1, 2, \ldots, \qquad (25.51)$$

we arrive at the *linear boundary value problems*

$$\frac{d^2u_m}{dx^2} - f'_2(x, y_m)u_m = v_m(x),$$

$$u_m(0) = u_m(1) = 0, \qquad m = 0, 1, 2, \ldots, \qquad (25.52)$$

for the generation of the Newton sequence $\{y_m(x)\}$. It is assumed that $y_0(x)$ satisfies the boundary conditions (25.47); actually, the problem may be restricted to the subspace $C''_{0,0}[0, 1]$ of $C''[0, 1]$ which consists of functions $y(x)$ satisfying (25.47), so that the boundary conditions would be fulfilled automatically.

The solution of linear boundary value problems such as (25.52) for second-order equations is by no means trivial. One possible approach employs the following integral equation technique. An inverse operator to the linear differential operator

$$L = \frac{d^2}{dx^2} \tag{25.53}$$

together with the boundary conditions (25.47) will be sought in the form of a linear integral operator G with kernel

$$G = G(x, t), \qquad 0\,x,\leq t\leq 1. \tag{25.54}$$

The kernel function (25.54) is called a *Green's function* after the English mathematician George Green (1793–1841). To find it, consider the simple linear boundary value problem

$$\frac{d^2y}{dx^2} = f(x),$$
$$y(0) = y(1) = 0. \tag{25.55}$$

The goal is now to express the solution of (25.55) in the form

$$y(x) = \int_0^1 G(x, t) f(t)\, dt \tag{25.56}$$

for $f \in C[0, 1]$. By straightforward integration,

$$y(x) = \int_0^x \int_0^s f(t)\, dt\, ds + c_1 x + c_2 \tag{25.57}$$

is the general solution of the differential equation $Ly = f$. Using the boundary conditions to determine the constants c_1, c_2 we find that

$$c_1 = -\int_0^1 \int_0^s f(t) \cdot dt\, ds,$$
$$c_2 = 0. \tag{25.58}$$

Substitution of (25.58) into (25.57) yields

$$y(x) = \int_0^x \int_0^s f(t)\, dt\, ds - \int_0^1 \int_0^s xf(t)\, dt\, ds. \tag{25.59}$$

The orders of integration in (25.59) can be interchanged on the basis of

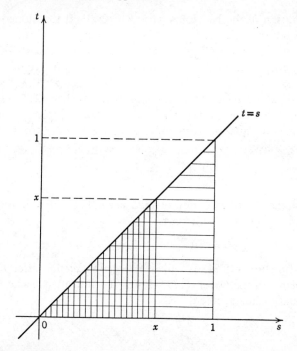

Figure 25.1   The regions of integration for the integrals (25.59).

inspection of Figure 25.1 to obtain

$$y(x) = \int_0^x \int_t^x f(t) \, ds \, dt - \int_0^1 \int_t^1 x f(t) \, ds \, dt$$

$$= \int_0^x \int_t^x f(t) \, ds \, dt - \int_0^x \int_t^1 x f(t) \, ds \, dt - \int_x^1 \int_t^1 x f(t) \, ds \, dt$$

$$= \int_0^x t(x - 1) f(t) \, dt + \int_x^1 x(t - 1) f(t) \, dt. \tag{25.60}$$

The last integrals of (25.60) are the desired result (25.56), with

$$G(x, t) = \begin{cases} t(x - 1), & 0 \le t \le x, \\ x(t - 1), & x \le t \le 1. \end{cases} \tag{25.61}$$

[Compare this with (9.29) and (9.30).]

Applying this result to (25.52), we get the sequence of linear integral equations of the Fredholm type and second kind

$$u_m(x) - \int_0^1 G(x, t) f_2'(t, y_m(t)) u_m(t) \, dt = w_m(x), \qquad 0 \le x \le 1, \tag{25.62}$$

where

$$w_m(x) = \int_0^1 G(x, t)\, v_m(t)\, dt, \qquad m = 0, 1, 2, \ldots. \tag{25.63}$$

As indicated in Chapter 1, Sections 9 and 10, it is possible to solve (25.62) by iteration under certain conditions, in particular, if

$$\|f'_{2m}\| = \max_{[0,1]} |f'_2(t, y_m(t))| \tag{25.64}$$

is sufficiently small.

For a specific example of a problem of the form (25.46) and (25.47), consider

$$\frac{d^2y}{dx^2} - (xy^2 - 1) = 0, \tag{25.65}$$

$$y(0) = y(1) = 0,$$

(see [24]). Here the linear problems (25.52) are

$$\frac{d^2u_m}{dx^2} - 2xy_m u_m = \frac{d^2y_m}{dx^2} + (xy_m{}^2 - 1), \tag{25.66}$$

$$u_m(0) = u_m(1) = 0, \qquad m = 0, 1, 2, \ldots.$$

For

$$y_0(x) = \tfrac{1}{2}x(1 - x), \qquad 0 \le x \le 1, \tag{25.67}$$

(25.66) becomes

$$\frac{d^2u_0}{dx^2} - x^2(1 - x)u_0 = \tfrac{1}{4}x^3(1 - x)^2, \tag{25.68}$$

$$u_0(0) = u_0(1) = 0.$$

The corresponding Fredholm integral equation (25.62) is

$$u_0(x) - \int_0^1 G(x, t)t^2(1 - t)\, u_0(t)\, dt = \frac{x^7}{42} - \frac{x^6}{15} + \frac{x^5}{20} - \frac{x}{140}, \qquad 0 \le x \le 1. \tag{25.69}$$

Using the norm

$$\|y\| = \max\{\|y\|_\infty, \|y'\|_\infty, \|y''\|_\infty\} \tag{25.70}$$

for $C''[0, 1]$, where

$$\|y\|_\infty = \max_{[0,1]} |y(x)| \tag{25.71}$$

is the norm for $C[0, 1]$, we have that the linear integral operator K from $C[0, 1]$ to $C''[0, 1]$ which has the kernel

$$K(x, t) = G(x, t)t^2(1 - t), \qquad 0 \le x, t \le 1, \tag{25.72}$$

has the bound

$$\|K\| \le \tfrac{4}{27} < 1, \tag{25.73}$$

so that the integral equation (25.69) has a unique solution $u_0(x)$ which can be found by the Neumann series, and

$$\|u_0\| \leq \frac{1}{1 - \|K\|} \left\| \frac{x^7}{42} - \frac{x^6}{15} + \frac{x^5}{20} - \frac{x}{140} \right\|_\infty \leq \frac{27}{23}\left(\frac{31}{420}\right) \simeq 0.087. \quad (25.74)$$

As

$$\|[P'(y_0)]^{-1}\| \leq \frac{1}{1 - \|K\|} = \frac{27}{23} \simeq 1.18, \quad (25.75)$$

we may take

$$B_0 = 1.18, \qquad \eta_0 = 0.087, \quad (25.76)$$

and since

$$P''(y) = -2xI_2, \quad (25.77)$$

is a constant bilinear operator from $C''[0, 1]$ into $C[0, 1]$,

$$\|P''(y)\| \leq K = 2 \quad (25.78)$$

on the whole space. Consequently,

$$h_0 = 0.20532 \leq \tfrac{1}{2}, \quad (25.79)$$

and Theorem 22.4 guarantees the existence of a solution $y^*(x)$ of the nonlinear boundary value problem (25.65) and the convergence of the Newton sequence $\{y_m(x)\}$ starting from $y_0(x)$ given by (25.67) to $y^*(x)$.

It would also be possible to use the Green's function (25.61) to obtain the nonlinear Fredholm integral equation

$$y(x) - \int_0^1 G(x, t)t\, y^2(t)\, dt = \tfrac{1}{2}x(1 - x), \qquad 0 \leq x \leq 1, \quad (25.80)$$

directly from (25.65). This illustrates the underlying principle that initial value problems for differential equations correspond to Volterra integral equations, and boundary value problems correspond to Fredholm integral equations [23].

More precisely, (25.80) is a nonlinear integral equation of *Hammerstein type* [see (15.15)], the general form of such equations being

$$P(y) = y(x) - \int_a^b K(x, t)\, g(t, y(t))\, dt = 0, \qquad a \leq x \leq b. \quad (25.81)$$

(The right-hand side of (25.80) can be absorbed into the nonlinear integral transform.) For

$$u_m(x) = y_{m+1}(x) - y_m(x),$$

$$v_m(x) = -y_m(x) + \int_a^b K(x, t)\, g(t, y_m(t))\, dt, \quad (25.82)$$

the Newton sequence for the solution of (25.81) would be generated by solving the sequence of linear integral equations of Fredholm type and second kind,

$$u_m(x) - \int_a^b K(x, t)\, g_2'(t, y_m(t))\, u_m(t)\, dt = v_m(x),$$

$$a \leq x \leq b, \quad m = 0, 1, 2, \ldots. \quad (25.83)$$

The nonlinear integral equations of *Urysohn type*,

$$P(y) = y(x) - \int_a^b K(x, t, y(t))\, dt = 0, \qquad a \leq x \leq b, \quad (25.84)$$

also lead to linear integral equations of Fredholm type and second kind for the generation of the Newton sequence, namely,

$$u_m(x) - \int_a^b K_3'(x, t, y_m(t))\, u_m(t)\, dt = v_m(x), \qquad a \leq x \leq b, \quad m = 0, 1, 2, \ldots,$$

$$(25.85)$$

where $u_m(x)$ is as defined in (25.82), and

$$v_m(x) = -y_m(x) + \int_a^b K(x, t, y_m(t))\, dt. \quad (25.86)$$

A general nonlinear integral equation of Fredholm type may be written in the form

$$P(y) \equiv \int_a^b K(x, t, y(x), y(t))\, dt = 0, \qquad a \leq x \leq b. \quad (25.87)$$

Here the generation of the Newton sequence depends on the solution of a sequence of linear integral equations of Fredholm type and *third kind*,

$$\phi_m(x)\, u_m(x) + \int_a^b K_4'(x, t, y_m(x), y_m(t))\, u_m(t)\, dt = v_m(x),$$

$$a \leq x \leq b, \quad m = 0, 1, 2, \ldots, \quad (25.88)$$

where

$$u_m(x) = y_{m+1}(x) - y_m(x),$$

$$\phi_m(x) = \int_a^b K_3'(x, t, y_m(x), y_m(t))\, dt, \quad (25.89)$$

$$v_m(x) = -\int_a^b K(x, t, y_m(x), y_m(t))\, dt, \qquad a \leq x \leq b, \quad m = 0, 1, 2, \ldots.$$

If it is possible to divide (25.88) by $\phi_m(x)$, then the resulting equation will again be of second kind, as in the previous example. The special case that $\phi_m(x) = 0, a \leq x \leq b$, is known as a linear integral equation of Fredholm type and *first kind*.

The nonlinear integral equation (13.1), which can be written as

$$y(x) - 1 - \frac{\varpi_0}{2} y(x) \int_0^1 \frac{x}{x+t} y(t) \, dt = 0, \qquad 0 \leq x \leq 1, \qquad (25.90)$$

can be considered to be of the form (25.87). From (25.89),

$$\phi_m(x) = 1 - \frac{\varpi_0}{2} \int_0^1 \frac{x}{x+t} y_m(t) \, dt, \qquad 0 \leq x \leq 1, \qquad (25.91)$$

so that the corresponding equation (25.88) will be of second kind if (25.91) does not vanish. The approximate solutions of (25.90) obtained previously indicate that (25.91) remains positive.

Additional examples and discussion of the application of Newton's method to the solution of nonlinear integral and ordinary differential equations may be found in the literature [23,24,25,26,27,28]. The book by H. T. Davis [29] also contains a number of interesting nonlinear integral and differential equations for which Newton's method could be investigated as an alternative solution procedure to the ones given.

To illustrate the application of Newton's method to nonlinear partial differential equations, consider the equation

$$\Delta u = u^2, \qquad (25.92)$$

where it is supposed that the boundary values

$$u(x, y) = b(x, y), \qquad (x, y) \in \partial G, \qquad (25.93)$$

are given on the boundary ∂G of a sufficiently regular region $G \subset R^2$ and a solution $u(x, y)$ of (25.92) is sought in the interior of G. This problem was mentioned previously in Section 11 [see (11.18), (11.19), and (11.20) as a particular example of (25.93)]. For the *partial differential operator*

$$P(u) = \Delta u - u \qquad (25.94)$$

from $C^2(G)$ to $C(G)$,

$$P'(u_0) = \Delta - 2u_0 I \qquad (25.95)$$

at $u_0 = u_0(x, y)$. In general, then, the Newton sequence would be generated by solving the *linear elliptic boundary value problems*

$$\Delta(u_{m+1} - u_m) - 2u_m(u_{m+1} - u_m) = -\Delta u_m + u_m{}^2,$$
$$u_{m+1}(x, y) = b(x, y), \qquad (x, y) \in \partial G, \qquad (25.96)$$

for $u_{m+1} = u_{m+1}(x, y)$, $m = 0, 1, 2, \ldots$, starting from some $u_0 = u_0(x, y)$. The problem (25.96) admits the immediate simplification to

$$\Delta u_{m+1} - 2u_m u_{m+1} = -u_m{}^2,$$
$$u_{m+1}(x, y) = b(x, y), \qquad (x, y) \in \partial G. \qquad (25.97)$$

Pohožaev [30] was able to show that, if $b(x, y) > 0$ and u_0 is taken to be the solution of the *Dirichlet problem* for the *Laplace equation*,

$$\Delta u_0 = 0,$$
$$u_0(x, y) = b(x, y), \qquad (x, y) \in \partial G, \tag{25.98}$$

then the sequence defined by (25.97) converges to a solution $u^* = u^*(x, y)$ of the problem of (25.92) and (25.93), and $u^*(x, y)$ is positive in the interior of G. The proof given by Pohožaev (which, incidentally, was carried out for the problem of (25.92) and (25.93) posed in R^3, but is valid also in R^2) makes use of a monotonicity argument and the *maximum principle* for the Dirichlet problem (25.98), namely,

$$\min_{(x,b) \in \partial G} b(x, y) \le u_0(x, y) \le \max_{(x,b) \in \partial G} b(x, y) \tag{25.99}$$

for $(x, y) \in G$ [6]. It is possible to establish a similar result by direct application of Theorem 22.4 and the use of an *integral equation technique*.

First of all, from (25.95),

$$P''(u) = -2I_2, \tag{25.100}$$

$$\|P''(u)\| \le 2 = K, \tag{25.101}$$

in the entire space $C^2(G)$. Thus to apply Theorem 22.4, satisfactory estimates for B_0, η_0 are needed. To find these, note that for

$$v_0(x, y) = u_1(x, y) - u_0(x, y) \tag{25.102}$$

the problem (25.96) becomes

$$\Delta v_0 - 2u_0 v_0 = u_0{}^2, \tag{25.103}$$
$$v_0(x, y) = 0, \qquad (x, y) \in \partial G,$$

assuming that $u_0(x, y)$ is the solution of the Dirichlet problem (25.98). The integral equation technique to be used is similar to that applied previously to the ordinary differential problem (25.52). We seek a *Green's function* $K(x, y; \xi, \eta)$ such that the solution of the *Poisson equation*

$$\Delta u = f(x, y) \tag{25.104}$$

with the *homogenous boundary value condition*

$$u(x, y) = 0, \qquad (x, y) \in \partial G \tag{25.105}$$

is given by the integral transform

$$u(x, y) = \int\int_G K(x, y; \xi, \eta) \, f(\xi, \eta) \, d\xi \, d\eta, \qquad (x, y) \in G. \tag{25.106}$$

Many such Green's functions are known; for example, if G is the square $0 \leq x, y \leq 1$, then [22]

$$K(x, y; \xi, \eta) = -\frac{4}{\pi^2} \sum_{\substack{j=1 \\ k=1}}^{\infty} \frac{\sin k\pi x \sin j\pi y \sin k\pi\xi \sin j\pi\eta}{j^2 + k^2}. \quad (25.107)$$

[The sign in (25.106) is customarily taken to be negative [22]; this has no bearing on the present discussion.]

By use of the Green's function, (25.103) becomes the linear integral equation of the Fredholm type and second kind

$$v_0(x, y) - 2\int\int_G K(x, y; \xi, \eta) \, u_0(\xi, \eta) \, v_0(\xi, \eta) \, d\xi \, d\eta = w_0(x, y), \quad (x, y) \in G,$$

$$(25.108)$$

where

$$w_0(x, y) = \int\int_G K(x, y; \xi, \eta) \, [u_0(\xi, \eta)]^2 \, d\xi \, d\eta. \quad (25.109)$$

If

$$2 \, \|K\| \cdot \|u_0\| < 1, \quad (25.110)$$

where $\|K\|$ is the bound of the *Green's integral operator in* $C(G)$ with kernel $K(x, y; \xi, \eta)$ and

$$\|u_0\| = \max_{(x,y) \in G} |u_0(x, y)|, \quad (25.111)$$

then the integral operator in (25.108) can be inverted and, by (9.15),

$$B_0 = \frac{1}{1 - 2\,\|K\| \cdot \|u_0\|} \geq \|[P'(u_0)]^{-1}\|, \quad (25.112)$$

while by (9.11),

$$\eta_0 = \frac{\|K\| \cdot \|u_0\|^2}{1 - 2\,\|K\| \cdot \|u_0\|} \geq \|v_0\| = \|u_1 - u_0\|. \quad (25.113)$$

Thus using (25.101), the convergence of Newton's method is assured if

$$h_0 = \frac{2\,\|K\| \cdot \|u_0\|^2}{(1 - 2\,\|K\| \cdot \|u_0\|)^2} \leq \frac{1}{2} \quad (25.114)$$

or

$$\frac{\|u_0\|}{1 - 2\,\|K\| \cdot \|u_0\|} \leq \frac{1}{2\,\|K\|} \sqrt{\|K\|}. \quad (25.115)$$

Since (25.115) can be written in the form

$$\|u_0\| \leq [2(\sqrt{\|K\|} + \|K\|)]^{-1}, \quad (25.116)$$

the size of $\|u_0\|$ determines the satisfaction of (25.114). By the maximum principle (25.99),

$$\|u_0\| \leq \max_{(x,y)\in\partial G} |b(x, y)|. \tag{25.117}$$

With the use of (25.116) and (25.117), we have the following theorem.

Theorem 25.1. If $u_0(x, y)$ satisfies the Dirichlet problem (25.98) and

$$\max_{(x,y)\in\partial G} |b(x, y)| \leq [2(\sqrt{\|K\|} + \|K\|)]^{-1}, \tag{25.118}$$

where $\|K\|$ is the bound of the Green's integral operator in G, then the Newton sequence defined by (25.97) converges to a solution $u^*(x, y)$ of the problem (25.92) and (25.93) in G.

This theorem does not restrict $b(x, y)$ to positive values. Once the value h_0 given by (25.114) is known, information about the rate of convergence of the Newton sequence and the regions of existence and uniqueness of the solution $u^*(x, y)$ are obtainable from the appropriate theorems in Section 22.

The method of *finite differences* applied to the foregoing problem, as in Section 11, yields a finite system of equations that can be handled in principle by the methods of Section 23. For the large linear systems which arise from partial differential equations, special techniques have been developed [10,11]. Greenspan has proved that, for the discretized version of the problem (25.92) and (25.93) in three dimensions, a theorem analogous to the one given by Pohožaev holds, and that the solutions of the finite system converge to the solution of the partial differential equation as the *mesh size h* approaches zero [10]. Greenspan also presents an interesting numerical example.

For the problem (11.18) and (11.20) in R^2, the corresponding finite difference equations are, by (11.26)

$$u_{ij}^2 + \frac{4}{h^2} u_{ij} - \frac{1}{h^2}(u_{i+1,j} + u_{i-1,j} + u_{i,j-1} + u_{i,j-1}) = 0,$$

$$i, j = 1, 2, \ldots, n - 1,$$

$$u_{i,0} = 2i^2h^2 - ih + 1, \qquad u_{n,i} = 2, \qquad i = 0, 1, \ldots, n,$$

$$u_{0,j} = 2j^2h^2 - jh + 1, \qquad u_{n,j} = 2, \qquad j = 0, 1, \ldots, n,$$

$$\tag{25.119}$$

where

$$h = \frac{1}{n}. \tag{25.120}$$

The system (25.119) reduces to

$$u_{11}^2 + 16u_{11} - 24 = 0 \tag{25.121}$$

for $h = \frac{1}{2}$ and the iterative formula obtained for the Newton sequence is

$$u_{11}^{(m+1)} = \frac{24 + (u_{11}^{(m)})^2}{2(8 + u_{11}^{(m)})}, \qquad m = 0, 1, 2, \ldots. \qquad (25.122)$$

Starting from

$$u_{11}^{(0)} = 1.5, \qquad -17.0, \qquad (25.123)$$

two iterations give

$$u_{11} = 1.3808, \, -17.3808, \qquad (25.124)$$

correct to four decimal places. For $h = \frac{1}{4}$, the general-purpose computer program described in Section 23 was used. For the initial approximations

$$u_{ij}^{(0)} = 1.0, \qquad i, j = 1, 2, 3, \qquad (25.125)$$

the program obtained the positive solutions, guaranteed accurate to seven decimal places, which are given in Table 25.1. Two iterations were required.

TABLE 25.1

Positive solutions of (25.119)

j / i	1	2	3
3	1.5167030	1.6258725	1.7642995
2	1.2097139	1.3877038	1.6258725
1	1.0259117	1.2097139	1.5167030

For the initial approximations

$$u_{ij}^{(0)} = -20.0, \qquad i, j = 1, 2, 3, \qquad (25.126)$$

the negative solutions given in Table 25.2 were obtained. These values are

TABLE 25.2

Negative solutions of (25.119)

j / i	1	2	3
3	−8.2014229	−15.3228386	−7.5527268
2	−16.6538945	−32.8628682	−15.3228386
1	−9.2167742	−16.6538945	−8.2014229

likewise accurate to seven decimal places, but 165 iterations were required for their calculation.

EXERCISES 25

1. Find an upper bound as a function of t for the linear integral operator K_m defined by (25.16). Hence obtain an upper bound for $\| [P'(y_0)]^{-1} \|$, where P is defined by (25.3).
2. Apply Theorem 22.4 to show that the initial value problem of (25.1) and (25.2) has a solution for t sufficiently small. *Hint.* Use the result of Problem 1 to obtain B_0, η_0. Estimate K with the use of (25.38).
3. Establish the identity

$$\int_0^t e^{-s^2} s^k \, ds = \frac{t^{k+1}}{k+1} e^{-t^2} + \frac{2}{k+1} \int_0^t e^{-s^2} s^{k+2} \, ds,$$

and use this result to obtain the power series expansion (25.24).
4. Verify that $y(x)$ as given by (25.60) is twice continuously differentiable if $f(x)$ is continuous, $0 \le x \le 1$. Calculate an upper bound for $\|G\|$ for the integral operator G from $C[0, 1]$ into $C''[0, 1]$ with kernel (25.61), taking

$$\|y\| = \|y\|_\infty + \|y'\|_\infty + \|y''\|_\infty$$

as the norm in $C''[0, 1]$, where $\| \ \|_\infty$ denotes the norm in $C[0, 1]$.
5. For the problem (25.65), find the first five nonvanishing terms of the power series for $y_1(x)$. *Hint.* Establish the identity

$$\int_0^1 G(x, t) t^n \, dt = \frac{x}{(m+1)(m+2)} (x^{m+1} - 1)$$

for $G(x, t)$ given by (25.61) and use the Neumann series solution for (25.69).

26. ACCESSIBILITY OF SOLUTIONS AND ERROR ESTIMATION

By analogy with Definition 12.2, a solution $x = x^*$ of the equation

$$P(x) = 0 \qquad (26.1)$$

is said to be *accessible from* x_0 *by Newton's method* if the Newton sequence $\{x_n\}$ starting from x_0 converges to x^*. For the *Newton iteration operator* F defined by

$$F(x) = x - [P'(x)]^{-1} P(x), \qquad (26.2)$$

we have

$$x^* = \lim_{m \to \infty} F^m(x_0) \qquad (26.3)$$

if x^* is accessible from x_0 by Newton's method. Theorems 22.1 through 22.3 give rather mild conditions under which the limit (26.3), if it exists, will be a solution of (26.1). As before, the *region of accessibility of* x^* *by Newton's method* is defined to be the set of all x_0 such that (26.3) holds.

The characterization of regions of accessibility of solutions of nonlinear operator equations by Newton's method is an interesting and difficult problem, even in the real scalar case [31]. An instructive experiment that can be performed with the aid of an electronic digital computer is to construct a *convergence map* for the zeros of a complex function $f(z)$ in a portion of the complex plane. To do this, take the rectangular region

$$R = \{z = x + iy : |x| \leq M, |y| \leq N\} \tag{26.4}$$

in the z-plane, and for

$$h = \frac{M}{m}, \qquad k = \frac{N}{n} \tag{26.5}$$

use the $(2m + 1)(2n + 1)$ mesh points

$$z_0(p, q) = ph + iqk,$$
$$p = -m, -m + 1, \ldots, -1, 0, 1, \ldots, m - 1, m, \tag{26.6}$$
$$q = n, n - 1, \ldots, 1, 0, -1, \ldots, -n + 1, -n,$$

as initial approximations to the zeros $z_1^*, z_2^*, \ldots$ of $f(z)$, that is, the solutions of the equation

$$f(z) = 0. \tag{26.7}$$

The computer is to be programmed to print the number r in the position corresponding to (p, q) if Newton's method converges to z_r^* starting from $z_0(p, q)$ or the number 0 if Newton's method fails to converge in a specified number of iterations to some preassigned accuracy.

Figure 26.1 shows a portion in the upper half-plane of the convergence map [32] for the polynomial

$$f(z) = z^3 + z^2 + 3z - 5 = (z - 1)(z + 1 - 2i)(z + 1 + 2i), \tag{26.8}$$

with

$$z_1^* = 1,$$
$$z_2^* = -1 + 2i, \tag{26.9}$$
$$z_3^* = -1 - 2i.$$

For the region shown,

$$h = k = 0.1,$$
$$p = -30, -29, \ldots, -1, 0, 1, \ldots, 29, 30, \tag{26.10}$$
$$q = 30, 29, \ldots, 1, 0.$$

The zeros z_1^*, z_2^* are enclosed in rectangles and the origin is circled in the figure. The map in the lower half-place is simply the reflection of the map in the upper half-plane with 2's and 3's interchanged, because of the symmetry in the location of the zeros of $f(z)$. It is interesting to note that Newton's method fails to converge at only two of the 1891 points shown in Figure 26.1.

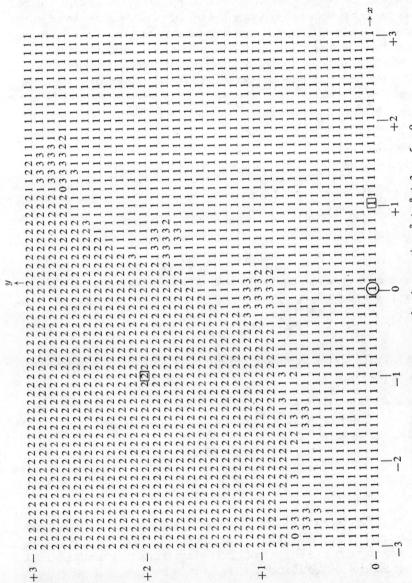

Figure 26.1 Convergence map for the equation $z^3 + z^2 + 3z - 5 = 0$.

It is not necessary to know the zeros of $f(z)$ in advance to construct a convergence map, as they may be numbered in the order in which they are obtained. For each initial value the iteration may be terminated when the convergence of the Newton sequence to a previously found zero can be guaranteed on the basis of Theorem 22.4 or one of the theorems to be given later in this section. Criteria for divergence may be taken to be inversion failure, the departure of some iterate from R or some larger region specified in advance, or if the number of iterations exceeds some preassigned value.

Two theorems will now be proved that provide information about the accessibility of a solution $x = x^*$ of (26.1) from points close to x^* or in a neighborhood of a point x_0 at which conditions for the convergence of the Newton sequence to x^* are satisfied.

Definition 26.1. A solution $x = x^*$ of (26.1) is said to be *simple* if $[P'(x^*)]^{-1}$ exists.

At a simple solution $x = x^*$, set

$$B^* = \| [P'(x^*)]^{-1} \|, \tag{26.11}$$

and assume that

$$\| P''(x) \| \leq K \tag{26.12}$$

for x in some open ball $U(x^*, r)$, $r > 0$. Without loss of generality, it may be supposed that

$$r > \frac{2}{B^* K}, \tag{26.13}$$

since K in (26.12) can be replaced, if necessary, by a larger number which satisfies (26.13).

Theorem 26.1. If x^* is a simple solution of (26.1) and $P''(x)$ is bounded in an open ball $U(x^*, r)$, $r > 0$, then x^* is accessible by Newton's method from any point x_0 such that

$$\| x_0 - x^* \| < \frac{1}{4B^* K}, \tag{26.14}$$

where K satisfies (26.12) and (26.13).

PROOF. This theorem will be established by showing that the hypotheses of Theorem 22.6 hold at x_0, which means that the Newton sequence, starting at x_0, will converge to a solution of (26.1) which is unique in a ball $U(x_0, \tilde{r}_0)$. Then it will be proved that $x^* \in U(x_0, \tilde{r}_0)$, which means that the Newton sequence starting at x_0 converges to x^*, since x^* is assumed to satisfy (26.1).

To obtain an estimate for B_0, note that (26.14) and (26.13) imply that

$\|x_0 - x^*\| < r$ and, thus,

$$\|P'(x_0) - P'(x^*)\| \leq K \|x_0 - x^*\| < \frac{1}{4B^*} < \frac{1}{B^*}. \tag{26.15}$$

By Theorem 10.1, $[P'(x_0)]^{-1}$ exists and is given by

$$[P'(x_0)]^{-1} = \sum_{m=0}^{\infty} \{I - [P'(x^*)]^{-1} P'(x_0)\}^n [P'(x^*)]^{-1}$$

$$= \sum_{n=0}^{\infty} ([P'(x^*)]^{-1}\{P'(x^*) - P'(x_0)\})^n [P'(x^*)]^{-1}. \tag{26.16}$$

From this and (26.15),

$$\|[P'(x_0)]^{-1}\| \leq \frac{B^*}{1 - B^*K \|x_0 - x^*\|} = B_0. \tag{26.17}$$

The estimation of η_0 is more involved. Using the fact that $P(x^*) = 0$,

$$\|x_1 - x_0\| = \|[P'(x_0)]^{-1} P(x_0)\|$$

$$= \|[P'(x_0)]^{-1}\{P(x_0) - P(x^*)\}\|$$

$$= \|[P'(x_0)]^{-1}\{P(x_0) - P(x^*) - P'(x^*)(x_0 - x^*)\}$$

$$+ [P'(x_0)]^{-1} P'(x^*)(x_0 - x^*)\|$$

$$\leq \|[P'(x_0)]^{-1}\| \cdot \|P(x_0) - P(x^*) - P'(x^*)(x_0 - x^*)\|$$

$$+ \|[P'(x_0)]^{-1} P'(x^*)\| \cdot \|x_0 - x^*\|. \tag{26.18}$$

By Taylor's theorem (20.45) and (26.12),

$$\|P(x_0) - P(x^*) - P'(x^*)(x_0 - x^*)\| \leq \tfrac{1}{2}K \|x_0 - x^*\|^2. \tag{26.19}$$

From (26.16),

$$[P'(x_0)]^{-1} P'(x^*) = \sum_{n=0}^{\infty} ([P'(x^*)]^{-1}\{P'(x^*) - P'(x_0)\})^n, \tag{26.20}$$

so that

$$\|[P'(x_0)]^{-1} P'(x^*)\| \leq (1 - B^*K \|x_0 - x^*\|)^{-1}. \tag{26.21}$$

Combining (26.18), (26.19), and (26.21), we find that

$$\|x_1 - x_0\| \leq \frac{1 + \tfrac{1}{2}B^*K \|x_0 - x^*\|}{1 - B^*K \|x_0 - x^*\|} \|x_0 - x^*\| = \eta_0. \tag{26.22}$$

Hence

$$h_0 = B_0 \eta_0 K = \frac{B^*K(1 + \tfrac{1}{2}B^*K \|x_0 - x^*\|)}{(1 - B^*K \|x_0 - x^*\|)^2} \|x_0 - x^*\|. \tag{26.23}$$

If (26.14) holds, then we can verify from (26.23) that

$$h_0 < \tfrac{1}{2}. \tag{26.24}$$

The ball of radius

$$\tilde{r}_0 = \frac{1 + \sqrt{1 - 2h_0\eta_0}}{h_0}$$

$$= \frac{1 - B^*K\,\|x_0 - x^*\| + \sqrt{1 - 4B^*K\,\|x_0 - x^*\|}}{B^*K} \qquad (26.25)$$

is contained in $U(x^*, r)$, since

$$\|x^* - x_0\| + \tilde{r}_0 = \frac{1 + \sqrt{1 - 4B^*K\,\|x_0 - x^*\|}}{B^*K} < \frac{2}{B^*K} < r. \quad (26.26)$$

Hence the hypotheses of Theorem 22.6 are satisfied. Since

$$2B^*K\,\|x_0 - x^*\| < 1$$

$$< 1 + \sqrt{1 - 4B^*K\,\|x_0 - x^*\|}, \qquad (26.27)$$

it follows at once that

$$\|x_0 - x^*\| < \frac{1 - B^*K\,\|x_0 - x^*\| + \sqrt{1 - 4B^*K\,\|x_0 - x^*\|}}{B^*K} = \tilde{r}_0, \quad (26.28)$$

which completes the proof.

Corollary 26.1. Under the hypotheses of Theorem 26.1, the simple solution x^* of (26.1) is unique in the open ball $U(x^*, 2/B^*K)$. This follows at once from (26.26).

For operators which have bounded second derivatives, Theorem 26.1 states that Newton's method will always converge to a simple solution x^* if we start sufficiently close to it, and the convergence will be *quadratic* ($h_0 < \frac{1}{2}$).

For *multiple* solutions x^*, that is, solutions of (26.1) for which $P'(x^*)$ is not invertible, the situation is more complicated [33]. Convergence to such solutions may be extremely slow, even in the case of scalar equations or finite systems. Furthermore, it is possible to construct examples where convergence is not obtained from initial points arbitrarily close to a multiple solution. For example, consider the system

$$\begin{aligned} x^2 + y^2 &= 0, \\ x^2 + 3y &= 0, \end{aligned} \qquad (26.29)$$

in R^2, which has the solution $(x^*, y^*) = (0, 0)$, with

$$P'\begin{pmatrix} x^* \\ y^* \end{pmatrix} = \begin{pmatrix} 0 & 0 \\ 0 & 3 \end{pmatrix}, \qquad (26.30)$$

a singular matrix. For an initial approximation of the form $(x_0, y_0) = (0, \epsilon)$, $\epsilon > 0$ arbitrarily small, we obtain the inconsistent linear system

$$2\epsilon(y_1 - y_0) = -\epsilon^2,$$
$$3(y_1 - y_0) = -3\epsilon, \tag{26.31}$$

by trying to use Newton's method.

Attention will now be turned to the question of accessibility of a solution x^* from points close to an initial point x_0 from which x^* is accessible by Newton's method. It will be convenient to denote such a solution of (26.1) by

$$x^* = x^*(x_0) \tag{26.32}$$

to indicate its accessibility from x_0.

Theorem 26.2. If the hypotheses of Theorem 22.6 are satisfied at $x = x_0$, then the Newton sequence will converge to $x^* = x^*(x_0)$ from any point x_0' such that

$$\|x_0' - x_0\| \leq \frac{1 - 2h_0}{4h_0} \eta_0. \tag{26.33}$$

PROOF. The reasoning will be carried out along the same lines as the proof of Theorem 26.1. For

$$\Delta_0 = \|x_0' - x_0\| \leq \frac{1 - 2h_0}{4h_0} \eta_0, \tag{26.34}$$

we have

$$\|P'(x_0') - P'(x_0)\| \leq K\Delta_0 < \frac{1}{B_0}, \tag{26.35}$$

so that $[P'(x')]^{-1}$ exists by Theorem 10.1 and

$$\|[P'(x_0')]^{-1}\| \leq \frac{B_0}{1 - B_0 K\Delta_0} = B_0'. \tag{26.36}$$

Also, as before,

$$[P'(x_0')]^{-1} P'(x_0) = \sum_{n=0}^{\infty} ([P'(x_0)]^{-1}\{P'(x_0) - P'(x_0')\})^n, \tag{26.37}$$

so that

$$\|[P'(x_0')]^{-1} P'(x_0)\| \leq \frac{1}{1 - B_0 K\Delta_0}. \tag{26.38}$$

This last inequality will be useful in obtaining an estimate for

$$\|x_1' - x_0'\| = \|[P'(x_0')]^{-1} P(x_0')\|, \tag{26.39}$$

since

$$[P'(x_0')]^{-1} P(x_0') = [P'(x_0')]^{-1} P'(x_0)\{[P'(x_0)]^{-1} P(x_0')\}. \tag{26.40}$$

The quantity in braces may be written as

$$[P'(x_0)]^{-1} P(x_0') = [P'(x_0)]^{-1}\{P(x_0') - P(x_0) - P'(x_0)(x_0' - x_0)\}$$
$$+ [P'(x_0)]^{-1} P(x_0) + (x_0' - x_0). \quad (26.41)$$

Use of the triangle inequality and (20.45) yields

$$\|[P'(x_0)]^{-1} P(x_0')\| \leq \tfrac{1}{2}B_0 K \|x_0' - x_0\| + \|[P'(x_0)]^{-1} P(x_0)\| + \|x_0' - x_0\|$$
$$\leq \eta_0 + \Delta_0 + \tfrac{1}{2}B_0 K \Delta_0^2. \quad (26.42)$$

Combining (26.38) and (26.42) gives

$$\|[P'(x_0')]^{-1} P(x_0')\| \leq \frac{\eta_0 + \Delta_0 + \tfrac{1}{2}B_0\Delta_0^2}{1 - B_0 K \Delta_0} = \eta_0'. \quad (26.43)$$

From (26.36) and (26.43),

$$h_0' = B_0' \eta_0' K = \frac{B_0 \eta_0 K + B_0 K \Delta_0 + \tfrac{1}{2}B_0^2 K^2 \Delta_0^2}{(1 - B_0 K \Delta_0)^2}. \quad (26.44)$$

If (26.33) is satisfied, then

$$h_0' \leq \frac{h_0 + (1 - 2h_0)/4 + (1 - 2h_0)^2/32}{[1 - (1 - 2h_0)/4]^2} = \frac{1}{2}. \quad (26.45)$$

From (26.44),

$$\tilde{r}_0' = \frac{1 + \sqrt{1 - 2h_0'}}{h_0'} \eta_0 = \frac{1 + \sqrt{1 - 2h_0'}}{B_0' K}$$

$$= \frac{1 - B_0 K \Delta_0 + \sqrt{1 - 2h_0 - 4B_0 K \Delta_0}}{B_0 K}. \quad (26.46)$$

Since

$$\tilde{r}_0' + \Delta_0 = \frac{1 + \sqrt{1 - 2h_0 - 4B_0 K \Delta_0}}{B_0 K} \leq \tilde{r}_0, \quad (26.47)$$

the hypotheses of Theorem 22.6 are satisfied at $x = x_0'$ and the Newton sequence starting from x_0' converges to a solution $x^* = x^*(x_0')$ of (26.1) which is unique in $U(x_0', \tilde{r}_0')$. As

$$U(x_0', \tilde{r}_0') \subset U(x_0, \tilde{r}_0) \quad (26.48)$$

by (26.47), $x^*(x_0') = x^*(x_0)$ and the theorem is proved, because $U(x_0, \tilde{r}_0)$ is the region of uniqueness of the solution $x^* = x^*(x_0)$ of (26.1).

Theorem 26.2 bears on the problem of *error estimation* in actual computation. This problem arises because the error bound (22.51) given in Theorem 22.5 is based on the assumption that all the terms of the Newton sequence are calculated exactly, which is seldom realized in practice. For various reasons to be considered in more detail later, we usually obtain a value x_1' at the first step of the computation which differs from the result x_1 called for by the theory. If it is possible to obtain an estimate

$$\Delta_1 \geq \|x_1' - x_1\| \tag{26.49}$$

for this computational error, then

$$\|x^* - x_1'\| \leq \|x^* - x_1\| + \|x_1' - x_1\| \leq 2h_0\eta_0 + \Delta_1 \tag{26.50}$$

by (22.51) and (26.49). At first glance the error bound (26.50) appears to be unsatisfactory, since the quantities η_0, h_0 are defined in terms of the unknown value $\|x_1 - x_0\|$. It is not necessary, however, to know $\|x_1 - x_0\|$ exactly if a satisfactory upper bound can be obtained for it; for example, if upper bounds

$$\delta_0 \geq \|P(x_0)\|, \qquad B_0 \geq \|[P'(x_0)]^{-1}\| \tag{26.51}$$

can be determined, we may take

$$\eta_0 = B_0\delta_0 \geq \|x_1 - x_0\| \tag{26.52}$$

and

$$h_0 = B_0\eta_0 K = B_0{}^2\delta_0 K. \tag{26.53}$$

If the conditions of Theorem 22.4 are satisfied for these values, then the error bound (26.50) is valid. Furthermore, if the hypotheses of Theorem 22.6 are satisfied at x_0 and

$$\Delta_1 \leq \frac{1 - 2h_1}{4h_1}\eta_1 = \frac{1}{1 - h_0}\frac{1 - 2h_0}{4h_0}\eta_0, \tag{26.54}$$

where (22.36) and (22.37) have been used, then Theorem 26.2 asserts that Newton's method will converge to $x^*(x_0)$ from x_1'. In terms of (26.52) and (26.53), (26.54) may be written as

$$\Delta_1 \leq \frac{1 - 2B_0{}^2\delta_0 K}{4B_0 K(1 - B_0{}^2\delta_0 K)}. \tag{26.55}$$

Probably the most satisfactory method for error estimation in connection with the use of Newton's method is to take the value x_1' obtained at the first step as a new value for x_0 and obtain estimates for B_0, η_0, K (if necessary), and Δ_1 at this point. The error bound (26.50) may then be applied. This process would be continued until a result x_1' of satisfactory accuracy is obtained or until an improvement in accuracy over that of x_0 is no longer indicated.

In connection with the estimation of η_0 it should be noted that in certain cases x_1 will be known exactly, but an approximate value x_1' may be more suitable for the subsequent calculations. This could be true if, for example, x_1 is expressed as an integral transform or as an infinite series, as was discussed previously in connection with the differential equation (25.19). Thus it might be possible to obtain η_0 (and also Δ_1) directly. In any case, it will be assumed that it is possible to evaluate $P(x_0)$ with sufficient accuracy to obtain an upper bound δ_0 for $\|P(x_0)\|$ as defined by (26.51). If P is finite-dimensional, then interval arithmetic could be used for this purpose [17,18].

A value for B_0 can be calculated on the basis of the assumption that two error bounds can be determined and are sufficiently small. First of all, $P'(x_0)$ is not calculated exactly, but we get a linear operator L_0 such that

$$\|P'(x_0) - L_0\| \leq \epsilon_0. \tag{26.56}$$

Secondly, L_0 is not inverted exactly, so we obtain L_1^{-1}, the exact inverse of an operator L_1 such that

$$\|L_0 - L_1\| \leq \epsilon_1. \tag{26.57}$$

The technique of obtaining ϵ_0 is sometimes called *forward error analysis* and ϵ_1 is found by *backward error analysis*. The latter procedure is now fairly standard in the finite-dimensional case [9, 12, 34, 35]. Knowing that

$$b_0 \geq \|L_1^{-1}\| \tag{26.58}$$

if

$$\epsilon = \epsilon_0 + \epsilon_1 \leq \frac{1}{b_0}, \tag{26.59}$$

then Theorem 10.1 guarantees the existence of $[P'(x_0)]^{-1}$ and

$$\|[P'(x_0)]^{-1}\| \leq \frac{b_0}{1 - \epsilon b_0} = B_0. \tag{26.60}$$

It is usually unnecessary to evaluate the upper bound

$$K \geq \|P''(x)\| \tag{26.61}$$

at each step of the iteration once a value has been found in a sufficiently large ball for which the hypotheses of Theorem 22.6 are satisfied, provided the error in the iterates remains bounded by (26.54) or (26.55). An automatic procedure for finding K in the finite-dimensional case was given in Section 24. More generally, suppose that P belongs to the class of *abstract analytic operators* which have the abstract power series expansion

$$P(x) = \sum_{n=0}^{\infty} \frac{1}{n!} P^{(n)}(x_0)(x - x_0)^n \tag{26.62}$$

at $x = x_0$. If P is an *abstract polynomial* (see Section 17), then the series (26.62) is finite and, hence, always convergent. In terms of the quantities

$$r = \|x - x_0\|, \qquad p_n \geq \|P^{(n)}(x_0)\|, \qquad n = 0, 1, 2, \ldots, \tag{26.63}$$

can define the *scalar majorant series* for P,

$$p(r) = \sum_{n=0}^{\infty} \frac{1}{n!} p_n r^n. \tag{26.64}$$

If (26.64) converges for $|r| < R$, then (26.62) converges uniformly in the ball $U(x_0, R)$, as does the series

$$P''(x) = \sum_{n=2}^{\infty} \frac{1}{(n-2)!} P^{(n)}(x_0)(x - x_0)^{n-2} \tag{26.65}$$

for the second derivative of P. From (26.64)

$$\|P''(x)\| \le p''(r), \qquad x \in U(x_0, r), \tag{26.66}$$

provided $r < R$. Thus we may take an upper bound for the second derivative of an abstract analytic operator P,

$$K(r) = p''(r), \qquad r < R, \tag{26.67}$$

or, if the dimensionless variable $s_0 = r/\eta_0$ introduced in Section 22 is preferred,

$$K(s_0) = p''(\eta_0 s_0), \qquad s_0 < \frac{R}{\eta_0}. \tag{26.68}$$

The error bound Δ_1 for $\|x_1' - x_1\|$ is essentially determined by the accuracy of calculation of

$$x_1 = x_0 - [P'(x_0)]^{-1} P(x_0) \tag{26.69}$$

or the solution of the linear equation

$$P'(x_0)(x_1 - x_0) = -P(x_0) \tag{26.70}$$

for x_1. In the finite-dimensional case this error analysis can be incorporated in the computer program used by employing interval arithmetic or other techniques [17, 34]. In general case, suppose that y_0 is obtained as the computed value for $P(x_0)$, with

$$\|P(x_0) - y_0\| < \delta. \tag{26.71}$$

Writing

$$x_1 - x_1' = [P'(x_0)]^{-1} P(x_0) - L_1^{-1} y_0$$

$$= \{[P'(x_0)]^{-1} - L_1^{-1}\} y_0 + [P'(x_0)]^{-1} \{P(x_0) - y_0\}, \tag{26.72}$$

we have from Theorem 10.1, (26.60), and (26.71)

$$\|x_1' - x_1\| \le \frac{\epsilon b_0^2 \|y_0\| + b_0 \delta}{1 - \epsilon b_0} = \Delta_1. \tag{26.73}$$

An important special case of this error estimation problem occurs if the linear equation (26.70) in an infinite-dimensional space X is approximated by a system of equations in a finite-dimensional subspace X_n of X. A typical example would be $X = R_\infty^\infty$, $X_n = R_\infty^n$. The space R_∞^∞ arises in connection with power series solutions of integral and differential equations, and if the vector of coefficients of such a series lies in this space, then the series will converge uniformly inside the unit interval. For Fourier series, we would be led to consider R_2^∞ in the same way.

The space R_∞^∞ has the interesting property that it can be written as the product of R_∞^n and itself; that is,

$$R_\infty^\infty = R_\infty^n \times R_\infty^\infty, \tag{26.74}$$

with the norm

$$\|x\| = \max \{\|x_n\|_n, \|x_\infty\|_\infty\}, \tag{26.75}$$

where $x = (x_n, x_\infty)$, $\|\ \|_n$ is the norm in R_∞^n, and $\|\ \|_\infty$ is the norm in R_∞^∞.

It will be supposed that

$$x_0 = (x_n^{(0)}, 0), \tag{26.76}$$

which may be considered to be a vector in R_∞^n for computational purposes. Furthermore, if

$$P(x_0) = (y_1^{(0)}, \ldots, y_n^{(0)}, y_{n+1}^{(0)}, \ldots), \tag{26.77}$$

take

$$y_0 = (y_1^{(0)}, \ldots, y_n^{(0)}, 0, 0, \ldots), \tag{26.78}$$

so that

$$\delta = \|(0, \ldots, 0, y_{n+1}^{(0)}, y_{n+2}^{(0)}, \ldots)\|. \tag{26.79}$$

The operator $P'(x_0)$ may be written as

$$P'(x_0) = \begin{pmatrix} L_{nn} & L_{n\infty} \\ L_{\infty n} & L_{\infty\infty} \end{pmatrix}, \tag{26.80}$$

according to the convention for linear operators in product spaces given in Section 7 [see (7.12)]. In this notation L_{nn} is an $n \times n$ matrix, $L_{n\infty}$ has n rows and an infinite number of columns, $L_{\infty n}$ has an infinite number of rows and n columns, and $L_{\infty\infty}$ is an infinite matrix. For L_0 take

$$L_0 = \begin{pmatrix} L_{nn} & 0 \\ 0 & I \end{pmatrix}, \tag{26.81}$$

where I is the infinite identity matrix. Thus

$$\epsilon_0 = \|P'(x_0) - L_0\| = \left\| \begin{pmatrix} 0 & L_{n\infty} \\ L_{\infty n} & I - L_{\infty\infty} \end{pmatrix} \right\|. \tag{26.82}$$

The inverse of L_0 can be obtained exactly if we can invert the finite matrix L_{nn}. Taking computational error into account, we have

$$L_1^{-1} = \begin{pmatrix} L_{nn}'^{-1} & 0 \\ 0 & I \end{pmatrix} \qquad (26.83)$$

instead, with

$$\epsilon_1 = \left\| \begin{pmatrix} L_{nn} - L_{nn}' & 0 \\ 0 & 0 \end{pmatrix} \right\|, \qquad (26.84)$$

which can be estimated by using the backward error analysis techniques for finite matrices. Of course, for all matrices $A = (a_{ij})$ under discussion, it is assumed that

$$\|A\| = \sup_{(i)} \sum_{j=1}^{\infty} |a_{ij}| \qquad (26.85)$$

is finite. Thus

$$x_1' = x_0 - L_1^{-1} y_0 \qquad (26.86)$$

may be obtained by calculations carried out in the finite-dimensional space R_∞^n. Assuming that ϵ, δ are sufficiently small, the error bound (26.73) holds for $\|x_2' - x_1\|$ and we can find B_0 from (26.60). As

$$\|P(x_0)\| \le \|y_0\| + \delta = \delta_0, \qquad (26.87)$$

we may take

$$\eta_0 = B_0 \delta_0 = \frac{b_0 \|y_0\| + b_0 \delta}{1 - \epsilon b_0}. \qquad (26.88)$$

Thus if a value for $K \ge \|P''(x)\|$ is known so that the conditions of Theorem 22.4 are satisfied at $x = x_0$, we have the error bound

$$\|x^* - x_1'\| \le K \left(\frac{b_0}{1 - \epsilon b_0} \right)^2 (\|y_0\| + \delta) + \frac{b_0}{1 - \epsilon b_0} (\epsilon b_0 \|y_0\| + \delta) \qquad (26.89)$$

by (26.50).

In some cases it will not be feasible to calculate x_1 or an approximation to it, but B_0, η_0, K can be found at $x = x_0$. In this event we may use the error bound for x_0,

$$\|x^* - x_0\| \le r_0 = \frac{1 - \sqrt{1 - 2h_0}}{h_0} \eta_0. \qquad (26.90)$$

In this way, error estimates may be obtained for initial approximations found by applying other soution techniques to the equation $P(x) = 0$, even though the procedure employed does not have an error estimation feature of its own.

A more extensive error analysis for Newton's method has been given by Lancaster [36].

EXERCISES 26

1. Determine the regions of accessibility by Newton's method of the solutions z_1, z_2 of the scalar quadratic equation
$$az^2 + bz + c = 0$$
($a \neq 0, b, c$ complex) in the complex plane. *Hint.* Transform the given equation into $w^2 - 1 = 0$ if $b^2 - 4ac \neq 0$, or into $w^2 = 0$ if $b^2 - 4ac = 0$, and determine the regions of accessibility for the solutions of these equations separately.

2. Construct a convergence map for
$$z^4 - 1 = 0$$
in the region
$$R = \{z = x + iy : |x| \leq 3, |y| \leq 3\}$$
in the complex plane, taking $h = k = 0.25$. *Hint.* By symmetry, it is sufficient to map the region
$$R' = \{z = x + iy : 0 \leq x \leq 3, 0 \leq y \leq 3\}$$
in the first quadrant.

3. If an electronic digital computer is available, repeat Problem 2 with $h = k = 0.1$.

4. Find an initial approximation (x_0, y_0) from which the Newton sequence will converge to the solution $(x^*, y^*) = (0, 0)$ of the system (26.29).

5. Extend the results of Theorem 26.1 and Corollary 26.1 to the closed ball $\bar{U}(x^*, 1/4B^*K)$. *Hint.* Take $\|x_0 - x^*\| = 1/4B^*K$ and apply Theorem 22.7.

6. Verify (26.46).

7. Obtain an error estimate for the approximate iterate
$$y_1' = t + \frac{t^3}{3} + \frac{2t^5}{15}$$
for the solution of the differential equation (25.19), T being small enough so that the hypotheses of Theorem 22.6 are satisfied at $y_0 = t$.

27.   VARIANTS OF NEWTON'S METHOD. INVERSION OF POWER SERIES

Many computational methods for the solution of nonlinear operator equations are related in some simple way to Newton's method or the method of successive substitutions considered in Chapter 2. For example, the calculation
$$x_{m+1}' = x_m' - L_m^{-1} y_m, \tag{27.1}$$
where L_m^{-1} is an approximation to $[P'(x_m')]^{-1}$ and y_m is close to $P(x_m')$, would be a *variant of Newton's method* arising from computational error in the generation of the exact Newton sequence, as considered in the previous section. Perhaps the variant of Newton's method of greatest interest is the *modified form of Newton's method,*
$$\hat{x}_{m+1} = \hat{x}_m - [P'(x_0)]^{-1} P(\hat{x}_m), \qquad \hat{x}_0 = x_0, \quad m = 0, 1, 2, \ldots. \tag{27.2}$$

For this procedure the labor of calculating $P'(x_0)$ and its inverse is done once and for all, after which we seek a fixed point of the *modified Newton iteration operator*

$$F(x) = x - [P'(x_0)]^{-1} P(x) \tag{27.3}$$

by successive substitutions. The following theorem, due to Kantorovič [3] gives the essential information concerning the convergence of this method.

Theorem 27.1.   If the hypotheses of Theorem 22.4 are satisfied at $x = x_0$ and

$$h_0 < \tfrac{1}{2}, \tag{27.4}$$

then the *modified Newton sequence* $\{\hat{x}_m\}$ defined by (27.2) converges to a solution $x = x^*$ of the equation

$$P(x) = 0. \tag{27.5}$$

with

$$\|x^* - \hat{x}_m\| \le 2h_0\eta_0(1 - \sqrt{1 - 2h_0})^{m-1}, \qquad m = 1, 2, \ldots. \tag{27.6}$$

PROOF.   The existence of x^* in the ball $\bar{U}(x_0, r_0)$,

$$r_0 = \frac{1 - \sqrt{1 - 2h_0}}{h_0}, \tag{27.7}$$

follows from Theorem 22.4. Now for

$$\hat{x}_1 = x_1 = x_0 - [P'(x_0)]^{-1} P(x_0), \tag{27.8}$$

suppose that

$$x \in \bar{U}(x_0, r_0) \cap \bar{U}(x^*, \|x_1 - x^*\|), \tag{27.9}$$

so that x satisfies the inequalities

$$\|x - x^*\| \le \|x_1 - x^*\|,$$
$$\|x - x_0\| \le \frac{1 - \sqrt{1 - 2h_0}}{h_0} \eta_0. \tag{27.10}$$

In particular, x_1 lies in the region (27.9). Define

$$\hat{x} = F(x), \tag{27.11}$$

where F is the iteration operator (27.3). Now, since

$$F'(x) = I - [P'(x_0)]^{-1} P'(x), \tag{27.12}$$

it follows that

$$F'(x_0) = 0. \tag{27.13}$$

Also by definition,

$$x^* = F(x^*). \tag{27.14}$$

Now for $\hat{x}$ given by (27.11),

$$\|\hat{x} - x^*\| = \|F(x) - F(x^*)\| \leq \sup_{\bar{x} \in L(x, x^*)} \|F'(\bar{x})\| \cdot \|x - x^*\|, \quad (27.15)$$

and another application of (20.45) yields

$$\sup_{\bar{x} \in L(x, x^*)} \|F'(\bar{x})\| = \sup_{\bar{x} \in L(x, x^*)} \|F'(\bar{x}) - F'(x_0)\| \leq \sup_{\substack{\tilde{x} \in L(x_0, \bar{x}) \\ \bar{x} \in L(x, x^*)}} \|F''(\tilde{x})\| \cdot \|\bar{x} - x_0\|. \quad (27.16)$$

From (27.12)

$$F''(x) = -[P'(x_0)]^{-1} P''(x), \quad (27.17)$$

and since

$$\|\bar{x} - x_0\| \leq \theta \|x^* - x_0\| + (1 - \theta) \|x - x_0\|$$

$$\leq \max \{\|x^* - x_0\|, \|x - x_0\|\} \leq r_0 \quad (27.18)$$

it follows from (27.15) through (27.17) that

$$\|\hat{x} - x^*\| \leq B_0 K r_0 \|x - x^*\| = (1 - \sqrt{1 - 2h_0}) \|x - x^*\|. \quad (27.19)$$

Thus $\hat{x} \in \bar{U}(x^*, \|x_1 - x^*\|)$ if $x \in \bar{U}(x^*, \|x_1 - x^*\|)$. To show that $\hat{x} \in \bar{U}(x_0, r_0)$, note that

$$\|\hat{x} - x_0\| \leq \|\hat{x} - x_1\| + \|x_1 - x_0\|$$

$$\leq \|F(x) - F(x_0) - F'(x_0)(x - x_0)\| + \eta_0$$

$$\leq \tfrac{1}{2} \sup_{\bar{x} \in L(x_0, x)} \|F''(\bar{x})\| \cdot \|x - x_0\|^2 + \eta_0$$

$$\leq \tfrac{1}{2} B_0 K r_0^2 + \eta_0 = r_0. \quad (27.20)$$

Hence $\hat{x}$ belongs to the region (27.9), which is mapped into itself by the operator F. The convergence of the modified Newton sequence $\{\hat{x}_m\}$ now follows immediately.

We have

$$\|\hat{x}_2 - x^*\| \leq (1 - \sqrt{1 - 2h_0}) \|\hat{x}_1 - x^*\| \quad (27.21)$$

by (27.8) and (27.19). By induction

$$\|\hat{x}_m - x^*\| \leq (1 - \sqrt{1 - 2h_0})^{m-1} \|\hat{x}_1 - x^*\|, \quad m = 2, 3, \ldots. \quad (27.22)$$

As

$$1 - \sqrt{1 - 2h_0} < 1 \quad (27.23)$$

for $h_0 < \tfrac{1}{2}$ and

$$\|\hat{x}_1 - x^*\| = \|x_1 - x^*\| \leq 2h_0 \eta_0 \quad (27.24)$$

by Theorem 22.5, (27.6) and the convergence of the modified form of Newton's method are established.

As an application of this procedure, we shall consider the characteristic value-vector problem for a linear operator L in a Banach space X. This problem is to find scalars λ, called *characteristic values*, and corresponding nonzero *characteristic vectors* x such that

$$Lx - \lambda x = 0. \tag{27.25}$$

(The Katzenjammer terms *eigenvalue* and *eigenvector* are used much more frequently than the historically and linguistically correct ones given in the foregoing.) This problem is easily formulated in the product space $Y = X \times \Lambda$ of X and its scalar field Λ [37]. For a linear functional ϕ and $y = (x, \lambda)$, we write

$$P(y) = \begin{pmatrix} Lx - \lambda x \\ \phi x - 1 \end{pmatrix} \tag{27.26}$$

The problem now can be solved by finding solutions of the equation

$$P(y) = 0, \tag{27.27}$$

for the condition

$$\phi x = 1 \tag{27.28}$$

ensures that $x \neq 0$.

At $y_0 = (x_0, \lambda_0)$ the derivative of P is

$$P'(y_0) = \begin{pmatrix} L - \lambda_0 I & -x_0 \\ \phi & 0 \end{pmatrix}, \tag{27.29}$$

and the second derivative of P is the constant bilinear operator

$$P''(y) = \begin{pmatrix} 0 & -I & -I & 0 \\ 0 & 0 & 0 & 0 \end{pmatrix} \tag{27.30}$$

If the norm

$$\|y\| = \max\{\|x\|_X, |\lambda|\} \tag{27.31}$$

is taken for Y, where $\|\ \|_X$ denotes the norm in X, then

$$\|P''(y)\| \leq 2 = K \tag{27.32}$$

in the whole space. It will be assumed that

$$L_0^{-1} = (L - \lambda_0 I)^{-1} \tag{27.33}$$

exists, which can happen only if λ_0 is *not* a characteristic value [see (27.25)]. For

$$u_0 = (I - \lambda_0 L)^{-1} x_0, \tag{27.34}$$

suppose that

$$\frac{1}{\xi_0} = \phi u_0 \neq 0. \tag{27.35}$$

Then $[P'(y_0)]^{-1}$ exists [37] and

$$[P'(y_0)]^{-1} = \begin{pmatrix} (I - \xi_0 u_0 \phi)L^{-1} & \xi_0 u_0 \\ -\xi_0 \phi L_0^{-1} & \xi_0 \end{pmatrix}. \tag{27.36}$$

From (27.36) it follows that we may take

$$B_0 = \max \{\|(I - \xi_0 u_0 \phi)L_0^{-1}\| + \|\xi_0 u_0\|, |\xi_0| (1 + \|\phi L_0^{-1}\|)\}, \tag{27.37}$$

where the norms are defined for vectors in X or operators on X. If x_0 is chosen so that

$$\phi x_0 = 1, \tag{27.38}$$

then

$$P(y_0) = \begin{pmatrix} Lx_0 - \lambda_0 x_0 \\ 0 \end{pmatrix} \tag{27.39}$$

and

$$\|P(y_0)\| = \|Lx_0 - \lambda_0 x_0\|_X. \tag{27.40}$$

The condition (27.4) for the convergence of the modified form of Newton's method becomes

$$\|Lx_0 - \lambda_0 x_0\|_X < \frac{1}{4B_0^2} \tag{27.41}$$

and the actual iteration has the form

$$x_{m+1} = x_m - (I - \xi_0 u_0 \phi)L_0^{-1}(Lx_m - \lambda_m x_m), \tag{27.42}$$

$$\lambda_{m+1} = \lambda_m + \xi_0 \phi L_0^{-1}(Lx_m - \lambda_m x_m), \qquad m = 0, 1, 2, \ldots, \tag{27.43}$$

it being assumed that (27.38) holds.

In R^m (or C^m), the condition $\phi x = 1$ could be taken to be that

$$\xi_k = 1 \tag{27.44}$$

for some k, $1 \le k \le x$, where $x = (\xi_1, \xi_2, \ldots, \xi_n)$. In a Hilbert space H we could select a fixed vector ϕ and define

$$\phi x = \langle x, \phi \rangle. \tag{27.45}$$

Thus in R_2^m (or C_2^m), (27.44) can be obtained by taking

$$\phi = (\delta_{1k}, \delta_{2k}, \ldots, \delta_{nk}), \tag{27.46}$$

where δ_{jk} is the Kronecker delta; for example, suppose that

$$L = \begin{pmatrix} 1 & 4 \\ 1 & 1 \end{pmatrix} \tag{27.47}$$

in R_2^2. Take

$$\phi = (0, 1), \qquad x_0 = (-1, 1). \tag{27.48}$$

As an initial approximation λ_0 to the characteristic value λ, use the *Rayleigh quotient*

$$\lambda_0 = \frac{\langle Lx_0, x_0 \rangle}{\langle x_0, x_0 \rangle} = -\frac{3}{2}. \tag{27.49}$$

Note that (27.49) gives the corresponding characteristic value if x_0 is a characteristic vector.

We have

$$L_0^{-1} = (L - \lambda_0 I)^{-1} = \begin{pmatrix} \frac{10}{9} & \frac{16}{9} \\ -\frac{4}{9} & \frac{10}{9} \end{pmatrix} \tag{27.50}$$

and

$$u_0 = L_0^{-1} x_0 = (-\tfrac{26}{9}, \tfrac{14}{9}). \tag{27.51}$$

It follows that

$$\xi_0 = \frac{1}{\phi u_0} = \frac{9}{14}. \tag{27.52}$$

For x_1, λ_1 formulas (27.42) and (27.43) take the especially simple form of

$$\begin{aligned} x_1 &= \xi_0 u_0, \\ \lambda_1 &= \lambda_0 + \xi_0. \end{aligned} \tag{27.53}$$

This gives

$$x_1 = (-\tfrac{13}{7}, 1), \qquad \lambda_1 = -\tfrac{6}{7}. \tag{27.54}$$

Continued iteration gives a sequence which converges to the solution $\lambda^* = -1$, $x^* = (-2, 1)$ of the characteristic value-vector problem for L given by (27.47).

The price paid for the computational simplicity of the modified form of Newton's method is a slower rate of convergence, as can be seen by comparison of (27.6) with (22.51) for $h_0 < \frac{1}{2}$. A study of the relative efficiency of Newton's method and the modified form has been made [7] to give a criterion as to which will attain a given accuracy with the least computation.

Other variants of Newton's method are considered by Bartle [38], namely,

$$x_{n+1} = x_n - [P'(z_n)]^{-1} P(x_n), \qquad n = 0, 1, \ldots, \tag{27.55}$$

where the points z_n are close to x_0, or

$$x_{n+1} = x_n - L_n^{-1} P(x_n), \tag{27.56}$$

where the L_n are linear operators near $P'(x_0)$. The process (27.56) is closely related to (27.1). The convergence rate given by Bartle's theorem is once again only geometric.

Consideration will now be given to the final topic of solution of the equation

$$P(x) = 0, \tag{27.57}$$

where P is an *abstract analytic operator* [see (26.62)] by the method of *inversion* (or *reversion*) *of power series*. In principle, this method will give an exact solution

$$x^* - x_0 = \sum_{n=1}^{\infty} A_n y_0{}^n \qquad (27.58)$$

of (27.57) when applicable, where y_0 is a known vector and the A_n are n-linear operators, $n = 1, 2, \ldots$. In practice, computation of only finite number of terms of (27.58) will be possible, resulting in an approximation

$$x_m = x_0 + \sum_{n=1}^{m} A_n y_0{}^n \qquad (27.59)$$

to x^*.

It will be supposed that (26.62) converges in a ball $U(x_0, r)$ large enough to contain x^*. Then

$$P(x^*) = \sum_{n=0}^{\infty} \frac{1}{n!} P^{(n)}(x_0)(x^* - x_0)^n = 0. \qquad (27.60)$$

If $[P'(x_0)]^{-1}$ exists, then from (27.60),

$$x^* - x_0 = -[P'(x_0)]^{-1} P(x_0) - [P'(x_0)]^{-1} \sum_{n=2}^{\infty} \frac{1}{m!} P^n(x_0)(x^* - x_0)^n. \qquad (27.61)$$

The relationship of this procedure to Newton's method is obvious since, if we approximate the series on the right-hand side of (27.61) by its first partial sum, we get

$$x_1 = x_0 - [P'(x_0)]^{-1} P(x_0). \qquad (27.62)$$

Write

$$y_0 = -[P'(x_0)]^{-1} P(x_0),$$

$$-[P'(x_0)]^{-1} \sum_{n=0}^{\infty} \frac{1}{n!} P^{(n)}(x_0)(x^* - x_0)^n = \sum_{n=2}^{\infty} B_n(x^* - x_0). \qquad (27.63)$$

Thus (27.61) becomes

$$x^* - x_0 = y_0 + \sum_{n=2}^{\infty} B_n(x^* - x_0)^n. \qquad (27.64)$$

By comparison with (27.59), we may take

$$A_1 = I, \qquad (27.65)$$

the identity operator. To obtain A_2 we substitute the value for $x^* - x_0$ given by (27.64) into the series on the right-hand side. (The validity of this procedure will be discussed later.) We get

$$x^* - x_0 = y_0 + B_2\left(y_0 + \sum_{n=2}^{\infty} B_n(x^* - x_0)^n\right)\left[y_0 + \sum_{n=2}^{\infty} B_n(x^* - x_0)^n\right]$$

$$= y_0 + B_2 y_0 y_0 + 2B_2 y_0 \sum_{n=2}^{\infty} B_n(x^* - x_0)^n + B_2\left[\sum_{n=2}^{\infty} B_n(x^* - x_0)^n\right]^2.$$

$$(27.66)$$

Thus

$$A_2 = B_2 = -\tfrac{1}{2}[P'(x_0)]^{-1} P''(x_0),\qquad (27.67)$$

using (27.63). Note that B_2 is symmetric, which was assumed in writing (27.66). Actually all of the operators A_n, B_n, $n = 2, 3, \ldots$ may be supposed to be symmetric. Thus the next approximation to x^* is

$$x_2 = x_0 + y_0 + B_2 y_0 y_0$$
$$= x_0 - [P'(x_0)]^{-1} P(x_0) - \tfrac{1}{2}[P'(x_0)]^{-1} P''(x_0)\{[P'(x_0)]^{-1} P(x_0)\}^2. \quad (27.68)$$

Iteration of (27.68) gives what is called *Cebyšev's method* for solving (27.57) [39,40]. This can be shown to converge faster than Newton's method under certain assumptions, but is considerably more complex from a computational standpoint.

We can continue in this manner to generate the successive multilinear operators A_3, A_4, $\ldots$, which are the coefficients of the series (27.59). These expressions quickly become very complicated, but a digital computer could be programmed to obtain them to any order desired.

To establish the validity of inversion of an abstract power series, use will be made of a corresponding result for scalar power series [41].

For

$$\eta \geq \|y_0\|,$$
$$\xi \geq \|x^* - x_0\|, \qquad (27.69)$$
$$b_n \geq \|B_n\|,$$

the expression

$$\xi = \eta + \sum_{n=2}^{\infty} b_n \xi^n \qquad (27.70)$$

corresponds to (27.64), the series on the right-hand side of (27.70) being a scalar majorant series for the right-hand side of (27.64). Suppose that (27.70) converges for $|\xi| < r$. Then the inversion technique previously described may be applied to (27.70) to obtain

$$\xi = \eta + \sum_{n=2}^{\infty} a_n \eta^n, \qquad (27.71)$$

where

$$a_n \geq 0, \qquad n = 2, 3, \ldots. \qquad (27.72)$$

Furthermore, the right-hand side of (27.71) will be a scalar majorant series for the infinite series in (27.58), that is,

$$a_n \geq \|A_n\|, \qquad n = 2, 3, \ldots, \tag{27.73}$$

with $\|A_1\| = 1$ by (27.65). Thus if (27.71) converges, then (27.58) converges and gives the solution x^* of (27.57). An error estimate for the approximation x_m defined by (27.59) is

$$\|x^* - x_m\| \leq \xi - \left(\eta + \sum_{n=2}^{m} a_n \eta^n\right) \tag{27.75}$$

which is a scalar majorant series for the remainder of the series (27.58) for x^*. It is assumed, of course, that $x^* \in U(x_0, r)$ from (27.71).

Although the method of inversion of series is quite complicated in general, there are some special cases in which the computation may be expressed in a fairly simple form. As an example, consider the abstract quadratic equation

$$x = y + \lambda Bxx, \tag{27.76}$$

where λ is a scalar parameter. The integral equations (13.1) and the corresponding discrete systems (13.35) considered in Section 13 are of this form, with

$$0 \leq \lambda = \frac{\varpi_0}{2} \leq \frac{1}{2}. \tag{27.77}$$

Writing the inverse series to (27.76) in the form

$$x = y + \sum_{n=2}^{\infty} \lambda^{n-1} A_n y^n, \tag{27.78}$$

corresponding to $x_0 = 0$, and setting

$$y^{(n-1)} = A_n y^n, \qquad n = 2, 3, \ldots, \tag{27.79}$$

we obtain the recursion relations

$$y^{(0)} = y,$$
$$y^{(m)} = \sum_{i=0}^{m-1} By^{(i)} y^{m-1-i}, \qquad m = 1, 2, \ldots \tag{27.80}$$

Hence a solution x^* of (27.76) is given by

$$x^* = \sum_{m=0}^{\infty} \lambda^m y^{(m)}, \tag{27.81}$$

provided that the indicated series converges.

The convergence analysis in this case is not difficult. The scalar majorant series for (27.76) is

$$\xi = \eta + \lambda b \xi^2 \tag{27.82}$$

for nonnegative λ. The inverse series to (27.82) is the binomial series for

$$\xi^* = \frac{1 - \sqrt{1 - 4\lambda b\eta}}{2\lambda b},$$ (27.83)

where

$$b \geq \|B\|, \qquad \eta \geq \|y\|.$$ (27.84)

This series converges for

$$\lambda < \frac{1}{4b\eta}.$$ (27.85)

Hence for the systems (13.35),

$$b = 0.69004017, \qquad \eta = 1.0,$$ (27.86)

and thus [42] the method of inversion of power series will work for

$$\lambda < \frac{1}{2.076016068} \leq 0.36230$$ (27.87)

or

$$\varpi_0 < 0.72460.$$ (27.88)

The values of $y^{(m)}$ for $m = 0, \dots, 31$ for the systems (13.35) are shown in Table 27.1 [43]. For $\varpi_0 = 0.1(0.1)0.7$, the results

$$x_{31} = \sum_{m=0}^{31} \lambda^m y^{(m)}$$ (27.89)

agree with those given in Table 13.3. For $\varpi_0 = 0.8$ the last few places are in error and convergence is not indicated for $\varpi_0 = 0.9$ or $\varpi_0 = 1.0$.

EXERCISES 27

1. Apply the contraction mapping principle (Theorem 12.2) to the convergence of the modified form of Newton's method, assuming that $\|P''(x)\| < K$ in $U(x_0, R)$, where

$$R = \frac{1}{B_0 K}.$$

Show that $h_0 \leq \frac{1}{4}$ is a sufficient condition for convergence. *Hint.* Establish that we may take $\theta(r) = B_0 K r$ and use the technique described in connection with Figure 13.1.

2. Verify (27.37) by direct operation on (27.29). Note that the identity operator in Y is

$$I = \begin{pmatrix} I_x & 0 \\ 0 & 1 \end{pmatrix},$$

where I_x is the identity operator in X.

3. For the iteration process (27.43), show that $\phi x_m = 1$ if $\phi x_0 = 1$, $m = 1, 2, 3, \dots$. *Hint.* Use (27.36) and (27.43) to prove that $\phi x_{m+1} = \phi x_m$, $m = 0, 1, 2, \dots$.

TABLE 27.1

The vectors $y^{(m)}$ of (27.80)

m	0	1	2	3	4	5	6	7
	1.0000000	0.0654181	0.0236329	0.0193681	0.022146	0.029560	0.043052	0.066314
	1.0000000	0.2115109	0.1225969	0.1126200	0.131804	0.176706	0.257437	0.396288
	1.0000000	0.3518652	0.2674465	0.2781746	0.344283	0.474138	0.700584	1.087248
	1.0000000	0.4649695	0.4169378	0.4772644	0.625095	0.891436	1.347276	2.123178
	1.0000000	0.5493060	0.5468476	0.6697716	0.917910	1.350926	2.088590	3.347400
	1.0000000	0.6094188	0.6489090	0.8323082	1.178982	1.778211	2.801163	4.555821
	1.0000000	0.6504428	0.7230433	0.9561371	1.385467	2.126450	3.396317	5.585826
	1.0000000	0.6764231	0.7718633	1.0402034	1.529110	2.373599	3.825805	6.339676
	1.0000000	0.6900402	0.7980290	1.0860590	1.608589	2.511976	4.068684	6.769640

m	8	9	10	11	12	13	14	15
	0.10621	0.17509	0.29512	0.50626	0.8809	1.5512	2.7589	4.9488
	0.63423	1.04480	1.75989	3.01729	5.2479	9.2371	16.4229	29.4500
	1.74888	2.89066	4.88050	8.38174	14.5970	25.7185	45.7617	82.1145
	3.45255	5.75243	9.77140	16.86107	29.4754	52.0935	92.9288	167.1090
	5.51391	9.28015	15.89164	27.60303	48.5177	86.1416	154.2664	278.3400
	7.59316	12.90271	22.27127	38.94337	68.8408	122.8234	220.8933	400.0380
	9.39594	16.08952	27.95402	49.15493	87.3151	156.4482	282.4225	513.1715
	10.73140	18.47505	32.24694	56.93136	101.4847	182.4029	330.1887	601.4541
	11.49875	19.85465	34.74382	61.47731	109.8056	197.7067	358.4578	653.8775

m	16	17	18	19	20	21	22	23
	8.942	16.264	29.749	54.686	100.96	187.22	348.36	650.34
	53.203	96.739	176.905	325.138	600.28	1112.76	2070.34	3864.82
	148.423	270.000	493.928	908.103	1677.04	3109.57	5786.76	10804.40
	302.607	551.341	1009.989	1859.123	3437.00	6378.89	11880.81	22199.46
	505.497	923.348	1695.259	3126.763	5790.82	10764.75	20078.59	37566.61
	728.901	1335.308	2458.012	4544.230	8433.84	15708.06	29350.32	55001.89
	937.823	1722.638	3178.662	5889.404	10952.26	20435.90	38248.39	71787.61
	1101.620	2027.606	3748.294	6956.517	12956.70	24210.39	45372.34	85261.95
	1199.182	2209.767	4089.423	7597.090	14162.61	26485.84	49675.21	93415.06

m	24	25	26	27	28	29	30	31
	1217.7	2286.7	4304.7	8123.7	15360	29104	55260	105084
	7236.2	13587.0	25576.7	48258.7	91252	172904	328240	624184
	20233.7	37997.0	71535.2	134993.5	255296	483780	918468	1746792
	41602.0	78173.7	147259.5	278036.7	526068	997332	1894240	3603920
	70484.7	132591.7	250021.5	472498.2	894772	1697684	3226820	6143516
	103350.2	194680.5	367559.7	695432.7	1318372	2503920	4763728	9077588
	135084.0	254795.0	481649.7	912340.7	1731436	3291732	6268452	11955520
	160620.2	303279.7	573866.2	1088016.5	2066612	3932128	7493668	14302604
	176097.5	332712.2	629928.0	1194964.2	2270924	4322984	8242324	15738344

4. Find another solution to the characteristic value-vector problem for the operator L defined by (27.47). *Hint.* Take ϕ orthogonal to $(-2, 1)$.

5. For the modified form of Newton's method, the *improvement factor* $\hat{\theta}_m$ corresponding to (22.53) is

$$\hat{\theta}_m = 2h_0^2(1 - \sqrt{1 - 2h_0})^{m-2}.$$

Compare the number of iterations N such that $\hat{\theta}_n \leq 10^{-k}$ with the values given in Table 22.1 for Newton's method, taking $h_0 = 0.1(0.1)0.4$, $k = 5, 10, 15$. If a digital computer is available, construct a table corresponding to Table 22.1, giving N for $h_0 = 0.01(0.01)0.49$, $k = 1(1)20$.

6. Give the inverse power series solution for

$$x = y + \lambda Txxx,$$

T being a trilinear operator in a Banach space X. What is the convergence condition corresponding to (27.85)?

REFERENCES AND NOTES

[1]   L. E. Dickson. *New First Course in the Theory of Equations.* Wiley, New York, 1939.

[2]   A. M. Ostrowski. *Solution of Equations and Systems of Equations.* Second ed. Academic, New York, 1966.

[3]   L. V. Kantorovič. Functional Analysis and Applied Mathematics. *Uspehi Mat. Nauk,* **3** (1948), 89–185 (Russian). Tr. by C. D. Benster, *Natl. Bur. Std. Rept. No. 1509,* Washington, 1952.

[4]   L. V. Kantorovič and G. P. Akilov. *Functional Analysis in Normed Spaces.* Russian ed. Moscow, 1959. Tr. by D. E. Brown. Pergamon, New York, 1964.

[5]   These computations were performed on the Burroughs B5500 computer of the University of Wisconsin Computing Center.

[6]   R. V. Churchill. *Introduction to Complex Variables and Applications.* McGraw-Hill, New York, 1948.

[7]   Lydia R. Lohr and L. B. Rall. Efficient Use of Newton's Method. *ICC Bull.,* **6** (1967), 99–103.

[8]   V. N. Faddeeva. *Computational Methods of Linear Algebra.* Tr. by C. D. Benster. Dover, New York, 1959.

[9]   B. Noble. *Applied Linear Algebra.* Preliminary ed. Prentice-Hall, Englewood Cliffs, New Jersey, 1966.

[10]   D. Greenspan. *Introductory Numerical Analysis of Elliptic Boundary Value Problems.* Harper and Row, New York, 1965.

[11]   R. S. Varga. *Matrix Iterative Analysis.* Prentice-Hall, Englewood Cliffs, New Jersey, 1962.

[12]   G. E. Forsythe and C. B. Moler. *Computer Solution of Linear Algebraic Systems.* Prentice-Hall, Englewood Cliffs, New Jersey, 1967.

[13]   These computations were performed on the CDC 3600 computer of the University of Wisconsin Computing Center.

[14] Julia H. Gray and L. B. Rall. NEWTON: A General Purpose Program for Solving Nonlinear Systems. *MRC Tech. Summary Rept. No. 790*, Mathematics Research Center, U.S. Army, University of Wisconsin, Madison, 1967.

[15] T. R. McCalla. *Introduction to Numerical Methods and FORTRAN Programming.* Wiley, New York, 1967.

[16] Julia H. Gray and A. Reiter. CODEX: Compiler of Differentiable Functions. *MRC Tech. Summary Rept. No. 791*, Mathematics Research Center, U.S. Army, University of Wisconsin, Madison, 1967.

[17] R. E. Moore. The Automatic Analysis and Control of Error in Digital Computing Based on the Use of Interval Numbers. In L. B. Rall (ed.), *Error in Digital Computation.* Vol. 1. Wiley, New York, 1965, 61–130.

[18] R. E. Moore. *Interval Analysis.* Prentice-Hall, Englewood Cliffs, New Jersey, 1966.

[19] A. Reiter. Interval Arithmetic Package (INTERVAL) for the CDC 1604 and the CDC 3600. *MRC Tech. Summary Rept. No. 794*, Mathematics Research Center, U.S. Army, University of Wisconsin, Madison, 1967.

[20] Max Morris and O. E. Brown. *Differential Equations.* Third ed. Prentice-Hall, Englewood Cliffs, New Jersey, 1952.

[21] K. Yosida. *Lectures on Differential and Integral Equations.* Interscience, New York, 1960.

[22] R. Courant and D. Hilbert. *Methods of Mathematical Physics.* Vol. 1. Interscience, New York, 1953.

[23] B. Noble. The Numerical Solution of Nonlinear Integral Equations and Related Topics. In P. M. Anselone (ed.), *Nonlinear Integral equations.* University of Wisconsin Press, Madison, 1964, 215–318.

[24] L. Collatz. *Functional Analysis and Numerical Mathematics.* German ed. Springer, Berlin, 1964. Tr. by Hansjörg Oser. English ed., Academic, New York, 1966.

[25] R. E. Bellman and R. E. Kalaba. *Quasilinearization and Nonlinear Boundary-Value Problems.* American Elsevier, New York, 1965.

[26] R. H. Moore. Newton's Method and Variations. In P. M. Anselone (ed.), *Nonlinear Integral Equations.* University of Wisconsin Press, Madison, 1964, 65–98.

[27] T. L. Saaty. *Modern Nonlinear Equations.* McGraw-Hill, New York, 1967.

[28] T. L. Saaty and Joseph Bram. *Nonlinear Mathematics.* McGraw-Hill, New York, 1965.

[29] H. T. Davis. *Introduction to Nonlinear Differential and Integral Equations.* Dover, New York, 1962.

[30] S. I. Pohožaev. The Dirichlet Problem for the Equation $\Delta u = u^2$. *Soviet Math. Dokl.*, **1** (1960), 1143–1146.

[31] Saul Gorn. Maximal Convergence Intervals and a Gibbs Type Phenomenon for Newton's Approximation Procedure. *Ann. of Math.*, **59** (1954), 463–476.

[32] These computations were performed on the CDC 1604 computer of the University of Wisconsin Computing Center.

[33] L. B. Rall. Convergence of the Newton Process to Multiple Solutions. *Numer. Math.*, **9** (1966), 23–37.

[34] E. L. Albasiny. Error in Digital Solution of Linear Problems. In L. B. Rall (ed.), *Error in Digital Computation.* Vol. 1. Wiley, New York, 1965.

[35]   J. H. Wilkinson. *Rounding Errors in Algebraic Processes.* Prentice-Hall, Englewood Cliffs, New Jersey, 1964.

[36]   P. Lancaster. Error Analysis for the Newton-Raphson Method. *Numer. Math.*, **9** (1966), 55–68.

[37]   P. M. Anselone and L. B. Rall. The Solution of Characteristic Value-Vector Problems by Newton's Method. *Numer. Math.*, **11** (1968), 38–45.

[38]   R. G. Bartle. Newton's Method in Banach Spaces. *Proc. Amer. Math. Soc.*, **6** (1955), 827–831.

[39]   M. I. Nečepurenko. On Čebyšev's Method for Functional Equations. *Usephi Mat. Nauk*, **9** (1954), 163–170 (Russian). Tr. by L. B. Rall. *MRC Tech. Summary Rept. No. 648*, Mathematics Research Center, U.S. Army, University of Wisconsin, Madison, 1966.

[40]   R. A. Šafiev. On Some Iterative Processes. *Ž. Vyčisl. Mat. i Mat. Fiz.*, **4** (1964), 139–143 (Russian). Tr. by L. B. Rall. *MRC Tech. Summary Rept. No. 649*, Mathematics Research Center, U.S. Army, University of Wisconsin, Madison, 1966.

[41]   Konrad Knopp. *Infinite Sequences and Series.* Tr. by Frederick Bagemihl. Dover, New York, 1956.

[42]   L. B. Rall. Quadratic Equations in Banach Spaces. *Rend. Circ. Mat. Palermo*, **10** (1961), 314–332.

[43]   These computations were performed on the LGP-30 computer of the Computing Center of Lamar State College of Technology.

The following papers are not cited in the text, but contain supplementary information of interest.

[44]   I. Fenyö. Über die Lösung der im Banachschen Raume definierten nichtlinearen Gleichungen. *Acta Math. Acad. Sci. Hungar.*, **5** (1954), 85–93 (Russian summary). This paper replaces the assumption of a bounded second derivative for the operator P by a uniform Lipschitz condition on the first derivative in the proof of the convergence of Newton's method.

[45]   A. A. Goldstein. On Newton's Method. *Numer. Math.*, **7** (1965), 391–393.

[46]   L. V. Kantorovič. The method of Successive Approximations for Functional Equations. *Acta Math.*, **71** (1939), 63–97.

[47]   V. E. Šamanskiĭ. A realization of the Newton Method on a Computer. *Ukrain. Mat. Ž.*, **18** (1966), 135–140 (Russian). Reviewed in *Mathematical Reviews*, **34**, No. 3781 (1967). The replacement of derivatives in the Jacobian matrix by finite difference quotients is suggested in this paper. This is probably of little advantage if a program for analytic differentiation is available and it complicates the error analysis considerably.

[48]   J. Schröder. Uber das Newtonsche Verfahren. *Arch. Rational Mech. Anal.* **1** (1957), 154–180. This paper treats the characteristic value-vector problem in some detail and contains a valuable bibliography.

ADDENDA to Chapter IV

Computational Solution of Nonlinear Operator Equations

The following papers establish the error bound

$$\|x^* - x_m\| \le \frac{\vartheta^{2^m}}{1 - \vartheta^{2^m}} \frac{2\sqrt{1 - 2h_0}}{h_0},$$

where $\vartheta = (1 - \sqrt{1 - 2h_0})/(1 + \sqrt{1 - 2h_0})$, which is generally sharper than (22.51), and in which equality holds for the scalar equation

$$P(x) \equiv \frac{h_0}{2\eta_0} x^2 - x + \eta_0 = 0 \qquad h_0 \leq \frac{1}{2} \quad,$$

for the Newton sequence starting with $x_0 = 0$.

[49] A. M. Ostrowski, La méthode de Newton dans les espaces de Banach, *C. R. Acad. Sci. Paris. 272* (1971), 1251–1253.
[50] W. B. Gragg and R. A. Tapia, Optimal error bounds for the Newton-Kantorovich theorem, *SIAM J. Numer. Anal. 11* (1974), 10–13.

An improved version of the computer program [14] has been written which uses interval matrix inversion to obtain rigorous upper bounds for η_0, B_0.

[51] Dennis Kuba and L. B. Rall, A UNIVAC 1108 Program for Obtaining Rigorous Error Estimates for Approximate Solutions of Systems of Equations, *MRC Tech. Summary Rept. No. 1168,* Mathematics Research Center, University of Wisconsin-Madison, 1972.

APPENDIX

AN ELEMENTARY INTRODUCTION TO THE LEBESGUE INTEGRAL

The sequence of continuous functions defined on $[0, 1]$ by

$$x_n(s) = s^n, \qquad n = 0, 1, 2, \ldots, \tag{A.1}$$

converges "pointwise" to the function

$$x^*(s) = \begin{cases} 0, & 0 \le s < 1, \\ 1, & s = 1. \end{cases} \tag{A.2}$$

This is simply to say that for each s in $[0, 1]$ the sequence of real numbers $\{s^n\}$ is convergent. If $0 \le s < 1$, then $s^n \to 0$, whereas for $s = 1$, $s^n \to 1$.

The norm on $C[0, 1]$ defined by $\|x\| = \max\limits_{s \in [0,1]} |x(s)|$ is sometimes called the *uniform norm* corresponding to the fact that a sequence of continuous functions on $[0, 1]$ which converges with respect to $\| \; \|$ is a *uniformly* convergent sequence of functions. The sequence of functions defined by (A.1) is *not uniformly* convergent. It is *not* a Cauchy sequence in $C[0, 1]$ with respect to $\| \; \|$. On the other hand, it *is* a Cauchy sequence with respect to the norm $\| \; \|_1$ defined by

$$\|x\|_1 = \int_0^1 |x(s)| \, ds. \tag{A.3}$$

This section was written by Professor Ramon E. Moore.

This is easily verified as follows:

$$\|x_{n+p}(s) - x_n(s)\|_1 = \int_0^1 |s^{n+p} - s^n|\, ds$$

$$= \left| \int_0^1 (s^{n+p} - s^n)\, ds \right|$$

$$= \left| \frac{1}{n+p+1} - \frac{1}{n+1} \right|.$$

Therefore $\|x_{n+p}(s) - x_n(s)\|_1 < \epsilon$ for every $p = 1, 2, \ldots$ if, for instance, $n > 1/\epsilon$.

Thus the linear space $C[0, 1]$ made into a normed linear space by $\|\ \|_1$ is *not* a Banach space since the Cauchy sequence defined by (A.1) does not converge to an element of $C[0, 1]$. The function defined by (A.2) is not continuous on $[0, 1]$.

We can extend the normed linear space of continuous functions on $[0,1]$ with the norm $\|\ \|_1$ to a Banach space by the *completion* process described in Section 5 of Chapter 1 [see (5.12) through (5.17)]. The resulting space is denoted by $L_1[0, 1]$. The integral (A.3) should be interpreted, however, as a *Lebesgue integral*.

There are sequences of functions in $C[0, 1]$ which are Cauchy sequences with respect to $\|\ \|_1$ [given by (A.3)] and which converge pointwise (that is, for each value of s in $[0, 1)$) to functions of the form

$$x^*(s) = \begin{cases} 0 & s \notin S, \\ 1 & s \in S, \end{cases} \tag{A.4}$$

where S is any finite set of points in $[0, 1]$.

In fact, by a diagonal process [see (5.16)] we can "construct" a sequence of continuous functions which converges with respect to $\|\ \|_1$ pointwise to (A.4), where S is the set of *all rational numbers* in $[0, 1]$. The resulting function is, however, *not integrable* in the sense of Riemann. The sums $\sum_{i=1}^n x^*(s_i)(s_{i+1} - s_i)$ do not converge as $n \to \infty$ to a unique number independent of the choice of the "mesh points" s_i. Clearly, if the s_i are all chosen as rational numbers, then the sums converge to 1, whereas if the s_i are all chosen as irrational numbers, then the sums converge to 0.

Other choices of s_i can produce numbers between 0 and 1 as limits of the sums for $n \to \infty$. The *Lebesgue integral* (a concept which we will now sketch briefly) of this function, on the other hand, *is* defined and is equal to 1.

The rational numbers in $[0, 1]$ form a *countable* set. In fact, the set of all positive rational numbers is countable. Consider the array

$$
\begin{array}{cccccc}
1/1 - 1/2 & 1/3 - 1/4 & \cdots & 1/n & \cdots \\
2/1 & 2/2 & 2/3 & 2/4 & \cdots & 2/n \\
\cdots & \cdots & \cdots & \cdots & \cdots \\
n/1 - n/2 & n/3 & n/4 & \cdots & n/n & \cdots
\end{array}
\tag{A.5}
$$

Every rational number m/n occurs at least once in this array. We can put the rational numbers and, *a fortiori*, the rationals in [0, 1] into 1-to-1 correspondence with the positive integers by counting along the dashed path indicated in (A.5) [rejecting duplicates, and, for that matter, rejecting rational numbers outside [0, 1] in the array (A.5)]. In this way we exhibit all the rational numbers in [0, 1] as a sequence $\{r_1, r_2, r_3, \ldots\}$.

Let ϵ be any positive real number. Suppose we put an open interval of width ϵ about the first rational number r_1, an interval of width $\epsilon/2$ about r_2, and so on. About r_n we put an open interval of width $\epsilon/2^{n-1}$. Then we have an open interval of *some* positive width about every rational number in [0, 1]. The sum of the widths of these open intervals is $\epsilon + \epsilon/2 + \epsilon/4 + \cdots + \epsilon/2^n + \cdots = 2\epsilon$. We conclude from all this that *all* rational numbers in [0, 1] can be covered with open intervals whose lengths add to an arbitrarily small positive number.

The *Lebesgue measure* of the set R of rational numbers in [0, 1] is *l.m.* $(R) = 0$. This means that the greatest lower bound of the total lengths of a set of open intervals containing the rational numbers is zero. The Lebesgue measure of the entire interval [0, 1] is $l.m.$ [0, 1] $= 1$ since that is clearly the greatest lower bound of the total length of any set of open intervals containing the whole set [0, 1].

Now if we remove the rational numbers from [0, 1] what is left is the set of irrational numbers in [0, 1]. The set of irrational numbers in [0, 1] thus has Lebesgue measure 1. In general, if an interval $[a, b]$ is represented as the union of two *disjoint* sets, $[a, b] = X_1 \cup X_2$ (with $X_1 \cap X_2$ empty), and if *l.m.* $(X_1) = L$, then *l.m.* $(|a, b|) = b - a = L + l.m.$ (X_2) and *l.m.* $(X_2) = b - a - L$.

With this much of an idea of the concept of Lebesgue measure we can indicate at least a rough idea of the Lebesgue integral. Suppose x is a bounded real-valued function on [0, 1] so that there are real numbers b and B such that $b < x(s) < B$ for all s in [0, 1].

Subdivide (partition) $[b, B]$ into N "subintervals" $[b_{i-1}, b_i]$ with

$$b = b_0 < b_1 < b_2 < \cdots < b_n = B.$$

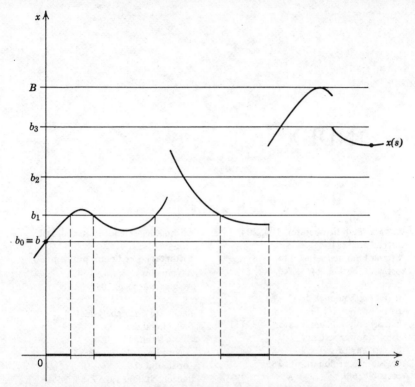

Figure A.1   A stage in the construction of a Lebesgue integral.

Define addition of two intervals by $[a, b] + [c, d] = [a + c, b + d]$ and multiplication of an interval by a nonnegative real number λ by $\lambda[a, b] = [\lambda a, \lambda b]$, and consider the sums

$$\sum_{i=1}^{n} [b_{i-1}, b_i] l.m. \{s: b_{i-1} \leq x(s) \leq b_i\}. \tag{A.6}$$

If the sums of the form (A.6) all converge to a single number independent of the choice of $\{b_i\}$ as long as $\max (b_i - b_{i-1}) \to 0$, $i = 1, 2, \ldots, n$, with $n \to \infty$, then this number is the value of the Lebesgue integral $\int_0^1 x(s) \, ds$ (see Figure A.1).

With this concept, we can see that the Lebesgue integral of the function on $[0, 1]$ given by

$$x(s) = \begin{cases} 0 & \text{for } s \text{ irrational,} \\ 1 & \text{for } s \text{ rational,} \end{cases} \tag{A.7}$$

is

$$\int_0^1 x(s) \, ds = 0. \tag{A.8}$$

INDEX

INDEX OF NOTATION

226